PERGAMON INTERNATI
of Science, Technology, Engineer
The 1000-volume original paperback li
industrial training and the enj
Publisher: Robert Maxwell, M.C.

LIVING EMBRYOS

Third Edition

Other Titles of Interest

ANDERSON

Embryology and Phylogeny in Annelids and Arthopods

COOPER

General Immunology

RAVEN

An Outline of Developmental Physiology. 3rd Edition

LIVING EMBRYOS

Third Edition

by

JACK COHEN

and

BRENDAN MASSEY

University of Birmingham

PERGAMON PRESS

OXFORD · NEW YORK · TORONTO · PARIS · FRANKFURT · SYDNEY

U.K.	Pergamon Press Ltd., Headington Hill Hall, Oxford OX3 0BW, England
U.S.A.	Pergamon Press Inc., Maxwell House, Fairview Park, Elmsford, New York 10523, U.S.A.
CANADA	Pergamon Press Canada Ltd., Suite 104, 150 Consumers Road, Willowdale, Ontario M2J 1P9, Canada
AUSTRALIA	Pergamon Press (Aust.) Pty. Ltd., P.O. Box 544, Potts Point, N.S.W. 2011, Australia
FRANCE	Pergamon Press SARL, 24 rue des Ecoles, 75240 Paris, Cedex 05, France
FEDERAL REPUBLIC OF GERMANY	Pergamon Press GmbH, 6242 Kronberg-Taunus, Hammerweg 6, Federal Republic of Germany

---

First edition 1963
Reprinted 1966
Second revised and enlarged edition 1967
Reprinted 1970
Third edition 1982

British Library Cataloguing in Publication Data

Cohen, Jack, *b. 1933*
Living embryos. — 3rd ed.
— (Pergamon international library).
1. Embryology
I. Title II. Massey, B
591.3′3 QL955 80-41106

ISBN 0-08-025926-X Hardcover
ISBN 0-08-025925-1 Flexicover

Printed in Hungary by Franklin Printing House

PREFACE

This, the 3rd edition of Living Embryos, has been prompted by the gratifying success of the first two. Obviously we are heartened by this success, but also conscious that in part at least it has been the format of the book that has contributed to its popularity. Accordingly, while trying to up-date this latest edition we have also attempted to retain the flavour of the previous two. In practice, this means that the superficial appearance of the text and illustrations will be reassuringly (we hope) familiar; it does not mean that the revision has been a superficial one. Every aspect of the book has received some attention, from the order of the sequence of the various sections, through re-writing most of the existing text to a greater or lesser degree, to completely new sections included for the first time; and many of the diagrams and photographs have been altered, too, in the hope of making them easier to understand.

Sections of the text have been reorganized in order to present more clearly the unfolding story of development; in particular, recent teaching experience has suggested that the tissue movements of gastrulation, often the stumbling block to an appreciation of development, can be more easily visualized if initially considered in a group such as the echinoderms —hence the reason for the apparently premature incursion into the mysteries of this phylum.

The "types" selected are *not* meant to be representative either of their phyla or of the wide range of development oddities found through the animal kingdom. The Nematoda have been chosen for the lesson of the eutelic creature, the ultimate in determinate development. A polychaete worm (*Pomatoceros*) has been chosen because it provides good practical material throughout the year and is cheap. It shows a complicated but readily understood cleavage pattern after post-fertilization interactions. Gastropod molluscs show the same kind of development and in addition the shell gland shows a post-cleavage induction. No excuse should be needed for inclusion of the arthropods, but their embryology is difficult to compare with other forms without considerably more space; the section has therefore been kept brief, but includes *Artemia* as an example since the embryology of this animal is easily demonstrated and eggs are available at all times from pet shops. The echinoderm serves as a good example of the indeterminate invertebrate, and is additionally both well-documented and good practical material at certain seasons. The tunicates have been included for two reasons: firstly, because they demonstrate radial cleavage in a determinate development and, secondly, to illustrate the dramatic resemblance between the determined areas of *Styela* and the frog fate map. This resemblance helps to introduce the vertebrates and underlines the importance of pre-cleavage events, for example the amphibian grey crescent.

The vertebrates have been emphasized, partly because man is a vertebrate and partly as a more detailed account was necessary from at least one phylum of the animal kingdom; the vertebrates were an obvious choice. The question of which organ systems to consider is also most important. Since the second edition our teaching in the areas of eye, ear, skin and urinogenital system embryology have prompted extensive re-writing of these sections (often to correct things that were plain *wrong* in the second edition!). New directions in embryological thinking, especially in the areas of reproductive theory and the maternal control of early embryology, have elicited completely new sections which we hope will stimulate readers to further investigation in texts which can afford a more generous cover than we can. Finally, diagrams have been added to this edition and most of the existing ones redrawn to some extent, mostly in response to student comment (or in response to their obvious, if unwitting, difficulty in understanding the old ones). However, the overall format is as of old and we hope it still pleases.

The aim of Living Embryos from its inception has been to treat embryology as a living subject, the very antithesis of the tedious plod through a set of slides, the pickled specimen, the list of someone else's photographs (which at best lead to boredom and at worst to a repetition of someone else's mistakes; and anyway it's much more fun to make your own mistakes from the start!). Wherever possible we have tried to encourage the student to refer to the living embryo, zygote, gamete or whatever, and the examples we have chosen to represent various diversions of the developmental theme are ones which are readily available to even the most impecunious devotees of embryology. It is worth emphasizing this last point (for those few of you who read the Preface!): all the examples chosen are *cheap*, reliable and easy to use as *living* developing material; all show something of the richness of the subject with particular clarity, be it the satisfaction of watching a *Pomatoceros* lay eggs or sperm, the precise cleavage of an echinoderm zygote, the magic of the first movements of a baby zebra fish, the beauty of the developing blood system of a 3-day chick, etc. Embryology seen like this is the most fascinating of sciences and deserves the widest audience.

Although Cohen started, Massey has been involved in the teaching for 15 years and has added his flavour throughout—we hope it is to your taste. In the preparation of this book in all its various manifestations, however, a large number of other people have cooperated as well and to thank them all personally would extend this preface beyond its already burgeoning length (and increase the price of the end product!). However, a special word of thanks is due to Chris Taplin, then a second year Zoology student in Birmingham, for his work on the existing diagrams and the very competent job he made of the many new ones—also for the patience he showed when we consistently mangled his efforts in an attempt to clarify our thoughts on what we wanted. And, of course, above all, thanks to those who made this edition necessary, the generation of students of all ages whose comments, criticism and advice are enshrined in it and whose successors, we hope, will be no less forthcoming.

Jack Cohen
Birmingham 1981 *Brendan Massey*

CONTENTS

LIST OF FIGURES

LIST OF TABLES

INTRODUCTION

The study of embryology is concerned primarily with the process by which the adult arises from the fertilized egg. It encompasses on the one hand those processes by which the genetic material expresses its message in material terms and on the other hand the maintenance of the adult organism, its progress into senility and its liability to suffer defects, such as tumours, which may result from developmental accident or design.

Until the beginning of this century embryology was essentially descriptive, but the development of experimental techniques has led to many unifying concepts. Not only the topology of embryological processes but also the causal relations between these processes should now be considered.

Let us first consider very briefly the history of the subject. The naive **preformationist** view that the egg or even the sperm was a miniature adult, requiring only to grow, was popular among natural philosophers (e.g. the animalculists). Some early drawings of human sperms purported to show a "homunculus" in the sperm head (Fig. 1). This view was current among scholastics until the end of the eighteenth century, when it became obvious that the facts could not be accounted for by such a simple story. Preformationist theories became very much more sophisticated and to account for the observed facts the process of **entelechy** was imported: the egg contained a "demon" which, during the process of development, organized the egg material in space to form the right kind of animal or plant. It is important to appreciate that neither the preformationist nor the entelechist considered the environment to be a necessary part of the process of development. During the nineteenth century workers began to be impressed by the apparent succession of stages in the life history of many creatures and those who were philosophers considered development in terms of a philosophical **evolution** (Aristotle had considered embryology in these terms): having exhausted the possibilities of each stage, the embryo *of necessity* progresses to the next stage. This viewpoint is held by many contemporary embryologists in a more or less sophisticated form, and human embryology still suffers considerably from the difficul-

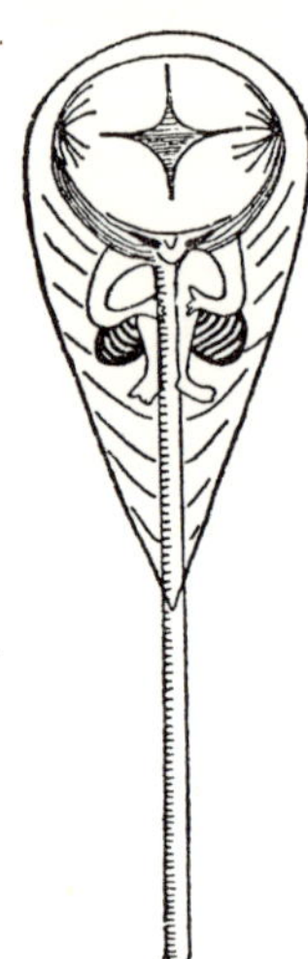

FIG. 1. *Early drawing of a human sperm by Hartsoeker (redrawn from Needham 1959).*

ties inherent in a consideration of stages (called **"horizons"**) instead of processes.

It may well be noted here that the view of DNA as "all of life; the rest is commentary" is a preformationist view with undertones of entelechy. The nucleic acids are an integral part of the structure of the egg and must not be considered as separate from it. They may play the part of the conductor, or indeed the composer, of the symphony of development; but the orchestra is absolutely necessary for its realization. Another analogy is that of a tape recording: the linear pattern of magnetic variation on the tape must interact with the complex and specific mechanism of the tape recorder for it to appear as a pattern of sound; similarly, the linear code of the DNA on the chromosomes must interact with the rest of the egg. The naive view that the DNA code is comparable to the sequence of frames making up a cine-film, requiring only to be expanded into space as the characters of the organism, shows its absurdity at first inspection. It has had an insidious effect, however, and still appears in many learned treatises as well as in newspaper popularizations. In such publications, the DNA is called the "blueprint"; if it *were* a blueprint, of course, the thalidomide tragedy would never have happened, for the embryos could have referred to their blueprints and corrected the developmental abnormalities.

Most contemporary biologists consider development in terms of **epigenesis;** this is the belief that the observed increase of complexity as an animal develops is due to *interaction* between its parts and often between these parts and the physical or chemical environment. The environment may contain **teratogens,** substances which divert the normal course of development into abnormal paths, like thalidomide. Abnormalities can also be produced by genetic differences, or by mechanical or chemical accidents to the egg or early embryo. It will be seen that the questions asked by entelechists or evolutionists can usually be answered in purely descriptive terms, whereas those of the epigeneticist require the experimental approach.

Throughout the study of embryology the time element must constantly be borne in mind. Even if certain stages in development may be described and recognized, the transition into and from these stages is a gradual process and indeed many events are occurring during what is apparently one stage. Models of stages of embryological development, which were very popular in the thirties, cannot show subtle physiological interactions, but are still useful for the modern biologist to get his anatomy straight. Films of the development of a great variety of forms are available on loan from several sources. Appropriate use of film material is often better than the living organisms in inexperienced hands; for one thing it always works! Viewing of a good film is also a useful preliminary to laboratory work. The student should gain some idea of the scale of the events he is to examine and, more important, should be able to select those organisms which are behaving normally and may be expected to continue to do so. Animated cartoon films have obvious dangers; the instructor should view them very critically himself before screening them for the students.

Reproduction involves the reproducing of parents from parents. When a cat has had kittens, reproduction has not yet occurred; only when the kittens in their turn have had kittens has the cycle been completed. **Embryology** is the central series of events whose scope extends from the gametes of one generation at least to the juveniles of the next. In fact,

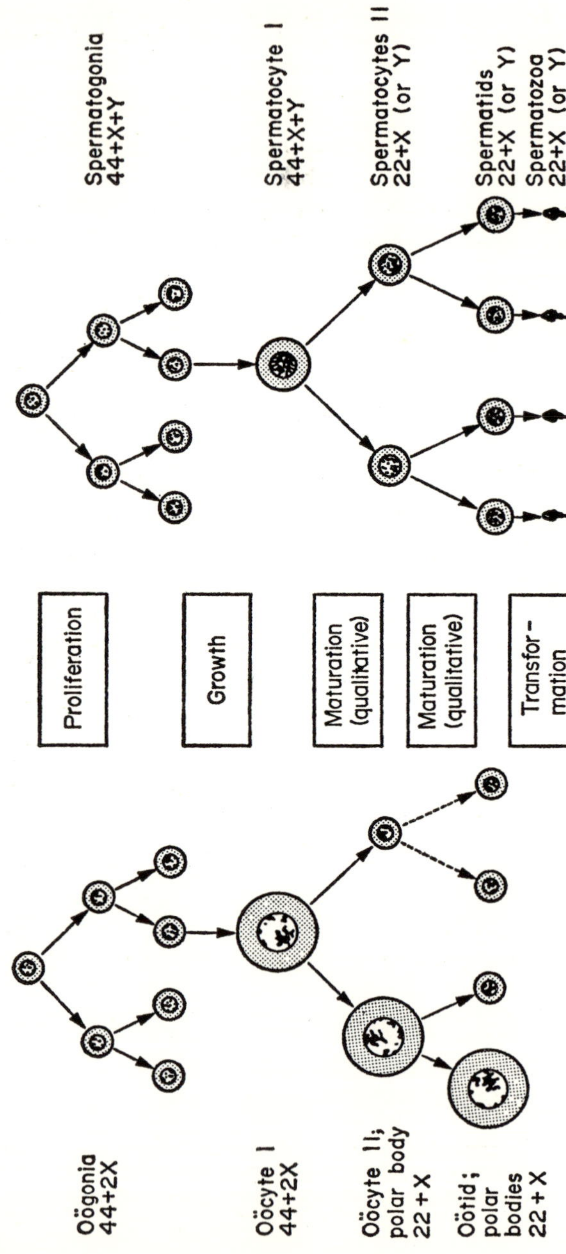

FIG. 2. *The production of eggs and sperms compared. This diagram shows the number of chromosomes for the human. Sex is determined by X or Y containing sperms; Y is male determining.*

many embryologists include the whole developmental process within embryology, i.e. not just the production of juveniles but also all the processes which enable them to produce their own gametes in the next generation. In this wide definition, embryology covers all the events necessary for reproduction.

Nearly all the organisms that we shall consider in this book reproduce sexually. The two sexes are *usually* housed in separate bodies and there is usually some specific behaviour associated with the transfer or fertilization of gametes. Such **sexual congress** often involves the animals in behaviour which is quite different from their normal **vegetative** activities, e.g. feeding, escaping from predators, remaining camouflaged. The reader should consult books on general reproductive topics for further information.

Such sexual organisms are commonly diploid and the production of gametes involves reduction division **(meiosis)** to produce haploid products. Again this is outside the scope of this book and the reader is referred to a textbook of genetics. In most animal species, the sexes are recognized primarily by the production of small, motile sperm or of large non-motile eggs; see Fig. 2 for a simple diagram of the development of both sperms and eggs. Many animals, especially many Protozoa, do not have any sexual process and in such forms there are no such **gametes;** here the embryology may be considered to commence with the division of the parent body. Many rotifers, insects and a few other forms normally reproduce parthenogenetically; that is to say, the eggs develop without fertilization by sperms. **Parthenogenesis** has been induced experimentally in many other groups and these parthenogenetic eggs and the embryology of the animals will not be considered separately, for in other respects they resemble their sexual relatives.

SPERMS

Spermatozoa (or **sperms**) of most animals (Fig. 3) move by means of a posterior flagellum, the tail, which is attached to the **head** by a **middle piece** (or **mid-piece**) containing mitochondria. The head of the sperm contains little other than the chromosomes and the **acrosome** which is used

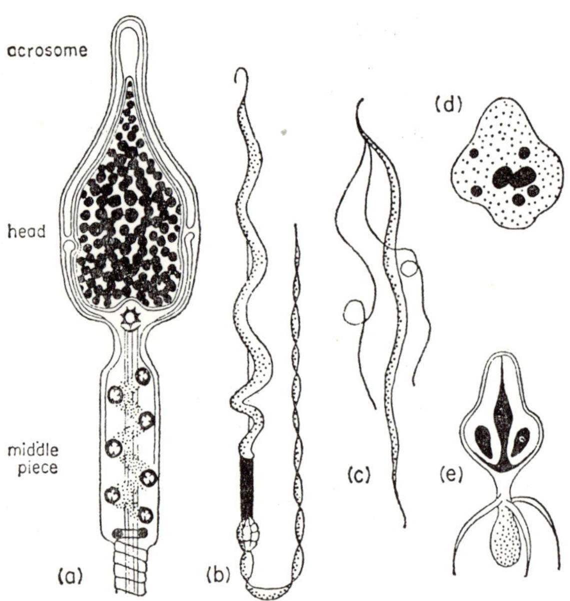

FIG. 3. *Sperms of various animals. (a) Detail of head and middle piece of bull sperm. (b) Sperm of a dogfish* (Squalas). *(c) Sperm of a planarian* (Dendrocoelum). *(d) Sperm of a nematode* (Ascaris). *(e) Sperm of a lobster* (Homarus). *Note that (d) and (e) are not flagellate.*

to penetrate the egg membrane. The acrosome of mammalian spermatozoa has been shown to contain enzymes, including hyaluronidase and proteases, which penetrate the membranes round the egg.

The sperms of many arthropods do not possess a true locomotory tail and some of them are very peculiar indeed; many are transferred in packets called **spermatophores.** Figure 3(e) shows the sperm of a lobster; crayfish sperm resemble tiny catherine wheels, with fibres extending tangentially from the edge of the disc; the sperms of some mites are rod-shaped, with a posterior bulge containing the nuclear material and minute grooves along their length which are thought to aid in locomotion. Some insects and spiders, on the other hand, have almost "normal" sperms.

Nematodes and some other Aschelminthes have sperms which are amoeboid in shape but do not seem to move very actively. Those of *Ascaris* may easily be seen in smears from the proximal oviduct of large females and a tiny rhythmic movement may be distinguished as they stick to the surface of eggs prior to entering them.

In order to produce the tiny heads characteristic of most sperms the chromosomes have to condense very greatly; their normal basic proteins, histones, are lost and replaced by protamines or other arginine-rich, very basic proteins, which bind the nucleic acids very tightly. They may be aligned very precisely; the use of polarizing microscopy of great delicacy has identified all of the chromosomes of a grasshopper arranged end-to-end in the long thin head of its sperm. However, most mammalian sperms have the chromosomes in no particular orientation; here, in addition to arginine-rich proteins, sulphur bridges between proteins give the head extra rigidity which enables it to penetrate the thick zona pellucida (p. 19) around the egg.

EGGS

The **egg** (Fig. 4) is usually more or less spherical and may be enclosed in a series of **membranes** which are often protective or contain nutrient materials. Eggs contain a nuclear apparatus in a very complex and highly organized **cytoplasm** whose components come from many sources; they also usually have some yolk. Most of the egg's substance has been accumulated during the process of **oogenesis;** ribosomes have been produced

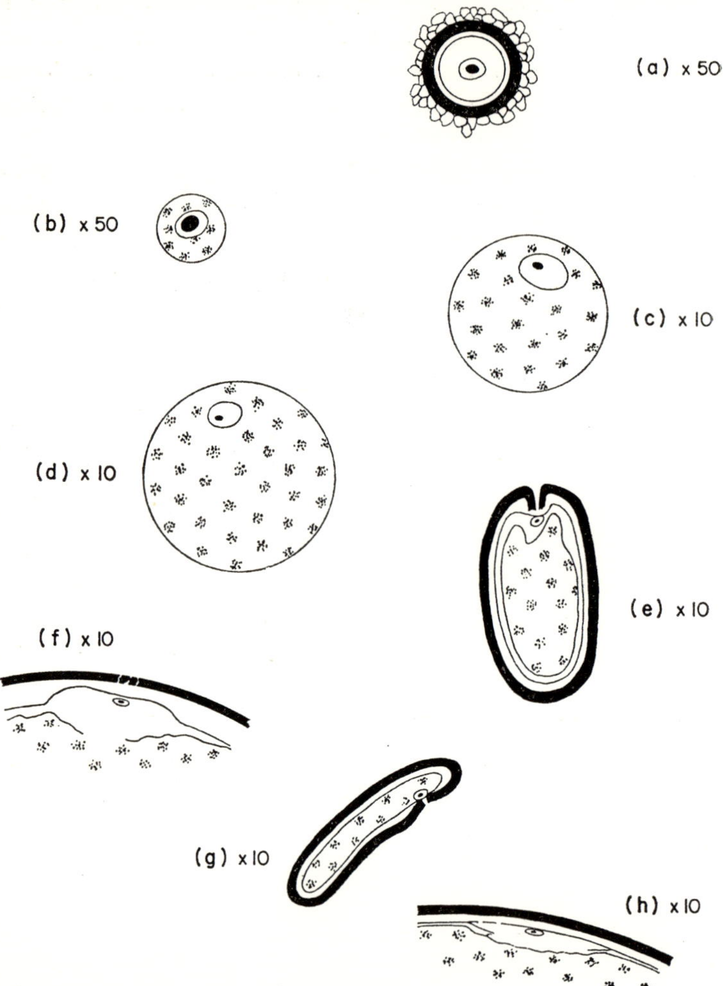

FIG. 4. *Egg sizes and structures.* (*a*) *Egg of a mammal* (*guinea pig*). (*b*) Pomatoceros. (*c*) *A newt* (Triturus). (*d*) Rana. (*e*) *Dwarf Cichlid* (Nannacara). (*f*) *Octopus.* (*g*) *Locust* (Schistocerca). (*h*) *Sparrow. The figures to the right indicate magnification.*

from multiplied nucleoli or have been injected into the egg from **nurse cells** during its development; mitochondria have been multiplied from a **mitochondrial cloud;** most important, a tremendous store of m-RNA has been accumulated which will serve for protein synthesis throughout cleavage and into later development.

The vast majority of eggs contain **yolk** which consists of fats, proteins and a substance called lecithin, all of which are distributed in the cytoplasm. These are the mother's usual contribution to the substance of the future embryo and probably provide most of the energy required for the developmental processes. However, she usually provides much more than this (see next section, p. 12).

Eggs without yolk are called **alecithal,** for example those of higher mammals. Eggs with yolk distributed evenly through the cytoplasm are called **homolecithal,** for example those of most marine invertebrates. Eggs with a moderate amount of yolk which displaces the nucleus towards one end are **mesolecithal,** for example the frog's egg. Those with a relatively enormous amount of yolk which makes the cytoplasm and nucleus form a separate disc, the **blastodisc,** are called **telolecithal;** examples are those of birds and sharks. The eggs of teleost fishes are difficult to define; previously called telolecithal, they may actually have less yolk than the frog egg but after fertilization the nucleus and yolk-free cytoplasm develop separately from the yolk, rather as in truly telolecithal eggs. The egg of insects is **centrolecithal** and the yolk occupies most of the volume of the lozenge-shaped capsule; the cytoplasm forms a thin layer over most of the surface and the nucleus is more or less deeply embedded in the yolk material (see Fig. 4).

Inside the protective and nutritive membranes is another membrane, associated with the egg cell membrane, which is called the **vitelline membrane.** This is of the utmost importance in subsequent development and is present in very nearly all eggs. The cytoplasm immediately beneath this vitelline membrane, especially in alecithal and homolecithal eggs, is an obviously different **cortex** whose granules are vital to the process of fertilization. Inside this cortex, there is usually a more fluid cytoplasm, perhaps with suspended yolk, and in this is the nucleus.

The egg when discharged from the ovary always has some additional organization, evidence for which is first shown by the direction of the spindles concerned in the nuclear divisions (meiosis) which produce the so-called **polar bodies** and render the egg haploid. These polar bodies are the "sister and aunt (or two cousins)" of the actual ovum resulting from meiosis (Fig. 2) and normally are discharged at that pole of the egg called the **animal pole** (or better the **cytoplasmic pole**), to which the egg nucleus is closest. In most animals, including the vertebrates, one polar body is extruded prior to fertilization and another at, or just after, fertilization (Figs. 2 and 5); that is because the egg usually waits for the sperm in the second metaphase of meiosis.

The existence of an animal pole implies, of course, a **vegetal pole** (or better, **yolky pole**) opposite to it. It will be seen that in mesolecithal and telolecithal eggs, because the nucleus has been displaced more or less towards the part of the surface called animal, the yolk occupies the vegetal part of the egg. There seems to be a tendency for the nucleus to lie approximately at the centre of the actual cytoplasm, disregarding inclusions like yolk. In a mesolecithal egg like that of most amphibians (Figs. 4 and 5) most of the yolk is in the vegetal hemisphere, so most of the "pure" cytoplasm is in the animal hemisphere. The nucleus is usually found about a third of the way from the animal to the vegetal poles. Of course, during the production of the polar bodies the nucleus approaches very close to the surface in most forms; as the spindle is always symmetrical and the polar bodies are so tiny it must do so if the cleavage plane is to divide the spindle. But, nevertheless, the "resting" position of the nucleus, for example the oocyte nucleus or the nuclear material just before first cleavage, is about the middle of the mass of the cytoplasm.

Very nearly all animals have a front end (usually the head), called **anterior,** and a rear end (often marked by the anus), called **posterior.** They thus have an antero-posterior **axis.** Most animals also usually have one surface down, called **ventral,** and one surface up, called **dorsal.** They therefore also have a dorso-ventral axis and definable left and right sides. (If you find this puzzling, consider *Amoeba* with no such axes; then *Hydra* with only one, and no left and right sides; then a fish, which has both axes and so has left and right sides.) Most eggs before fertilization

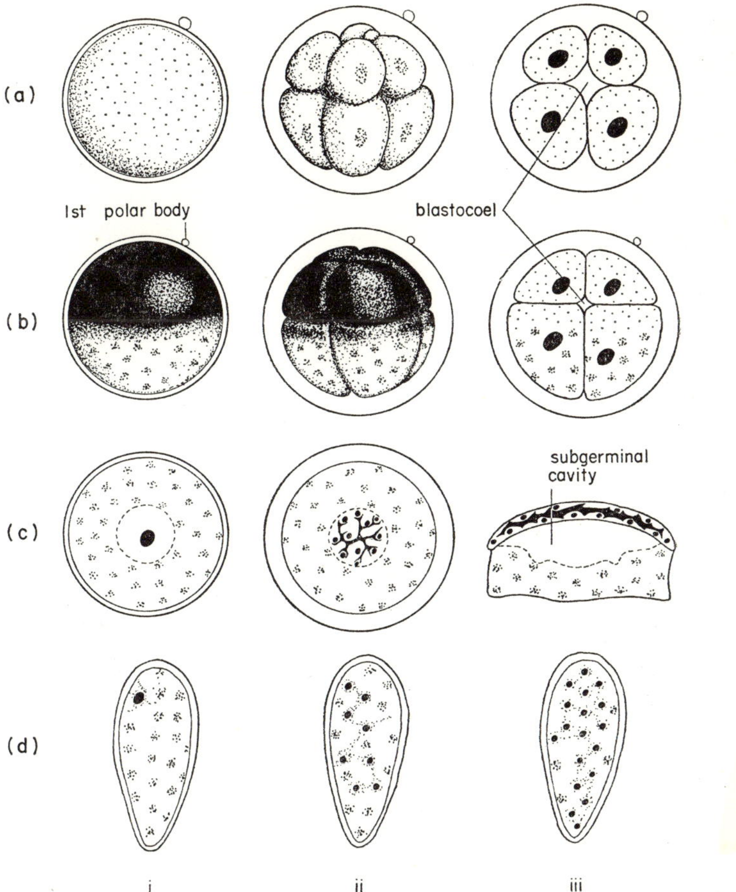

FIG. 5. *Cleavage in various eggs. (a) Homolecithal egg of a starfish. (b) Mesolecithal egg of a frog. (c) Telolecithal egg of a fish* (Platypoecilus). *(d) Centrolecithal egg of an insect. (i), (ii), (iii) show unfertilized egg, 8-cell stage, and section of 8-cell stage, except in (d), where sections are represented throughout and only the nuclei cleave*

only have one such axis, the animal–vegetal: only by cutting an egg along "lines of latitude" would the slices all be different. If it was cut along "lines of longitude" the segments (like those of an orange) would all appear the same. Many eggs wait for the arrival of the sperm to mark a point on the surface as 0° longitude (the "position of Greenwich"); for example some frogs use the position of sperm entry to define bilateral symmetry. Some animals define their bilateral symmetry while the egg is still in the ovary, for example the gastropod molluscs; in this case, the nutritive cells which pass material into the developing oocyte are arranged in a pattern around the egg and the orientation of the future embryo is determined by this, independent of where the sperm may actually enter. The egg of the guppy is another example of this; in this case, the area of the blastodisc which will lie at the head of the young fish is already thickened before fertilization and the embryo will always develop with its head in this region, wherever the sperm enters. Indeed, many sperms normally do enter the guppy egg, all in different places; a normally orientated embryo still results. The eggs of other animals, for example *Hydra*, have a special region at one part of the surface which is particularly "attractive" to sperms and in such cases as this the position of sperm entry may be more or less definite. Sometimes, as in many fishes and insects (see Fig. 4), the unfertilized egg has a thick shell called a **chorion** (not to be confused with the embryonic membrane with the same name; this one is probably always secreted before fertilization) with only one pore, the **micropyle,** through which the sperm enters.

Prior to fertilization the egg may also contain various pigments, which may be arranged either homogenously (as in the worm, *Pomatoceros triqueter*), or in an animal-to-vegetal succession (as in *Styela*, a tunicate).

FURTHER CONTRIBUTIONS BY THE MOTHER

The mother nearly always provides much more than yolk and chromosomes to the future of her offspring. Indeed, it is usual for her to provide all of the instructions for early development, as well as the wherewithal to achieve it. Father only contributes a genotype (usually differing very little from that of the mother) which is not involved in development until the phyletic stage (p. 68). Because of the great involvement of maternal resources in a variety of ways, the embryos "choice of a mother" determines many of its advantages/disadvantages compared with its contemporaries. We may classify mother's contribution to embryonic survival into five different categories and these relate to her genotype on the one hand and **privilege** on the other; both contribute to the inheritance of the embryo.

(1) The mother provides one set of chromosomes to the zygote nuclear chromosomal complement. She also provides the egg with an architecture which will take it through early development to the point (phyletic stage, p. 68) where these nuclear genes can begin to determine the course of its further development.

(2) Direct effects of the maternal nuclear genotype on embryonic development have been demonstrated. Examples are: the direction of coiling of gastropod shells, whose coiling sense depends on mother's genetics; grandchildlessness (p. 53) in *Drosophila*.

(3) There are many informational DNA molecules other than those associated with the mother's nucleus and these may be passed to offspring. They range from chloroplast and mitochondrial genomes (which may determine sensitivity to antibiotics), through *Drosophila* A-bodies in the eggs (*Rickettsia*, necessary for development) to symbionts like the Protozoa of termites or cows (necessary for digestion of plant material). Disease caught from the mother could, of course, have a negative effect on the development of embryos ("hereditary" syphilis, rabbit fleas).

(4) The mother's phenotype will also affect the start in life that she can give the embryos. Although this depends to some extent on her genotype, so many accidents are involved during her maturation that it is very difficult to tie any individual property of a good mother solely to her genetics. A competent mother will make "profit" on her metabolic transactions with the environment and "invest" these in the offspring as yolk, milk or care. Mother's appearance can, of course, considerably affect the available choice of fathers for the offspring!

(5) Some mothers may accidentally have properties not at all related to their genetics, which will contribute good or ill to their embryos. Such effects range from mothers who expose embryos to nicotine or thalidomide, to bottle or breast feeding or even the choice of a language.

The above very wide consideration of the maternal contribution to embryonic inheritance should help counter the belief that only the DNA, donated equally by both parents, is important for development; the yolk is as much a part of the inheritance of the embryo as is the DNA component.

NUMBERS OF GAMETES

The number of eggs produced varies enormously from one species to another; for example the cod lays about 20,000,000 each year and the oyster about 40,000,000, but birds rarely lay more than 20 eggs in a year. Nevertheless, two cod parents, oyster parents or bird parents normally reproduce two parents, on average, in the next generation. (If they did not, they would die out or there would be a population explosion.) The millions of eggs produced by some organisms each have very little parental care, whereas when few offspring are produced there is frequently enormous parental investment in each one. It is found that organisms which adopt the first strategy, that of prolific egg production, also tend as adults to be short-lived, versatile, environmental pioneers; those adopting the second strategy (few, well-tended eggs) tend to be long-lived animals, very closely adapted to their environment, with a secure niche in a complicated and stable ecology. There are, of course, many organisms in the middle range between these two extremes, but it is convenient to consider reproductive strategies as ranging from the r end (r = theoretical maximum rate of population increase) to the K end (K = carrying capacity of the environment). Within any group of animals of course, there are r and K adapted members; for example, although mammals are normally thought to be K-adapted, the mouse is r-adapted for a mammal (about 25 offspring per year); on the other hand the rhinoceros is strongly K-adapted (1 offspring produced at a time, with a gap of several years between them).

When we turn to the number of sperms produced by animals, again we find variation, but in general the great majority of male animals produce very large numbers of sperms to achieve fertilization. For example, man can produce 350,000,000 sperms in one normal ejaculation; obviously the *r* and *K* story used to explain variations in egg number cannot apply here.

It was the contention of old embryologists that the "profligacy of nature", almost universally associated with sperm production, was a left-over from ancient external fertilization, which they assumed to be grossly inefficient. It has also been suggested that, at least in animals in which fertilization is internal, the tortuousness of the female tract may require a heavy "infection" of sperms to enable them to diffuse past barriers designed to keep bacteria away from the eggs. A recent suggestion is more interesting; this is that the process of spermatogenesis is not perfect and some processes (notably genetical crossing over) may "go wrong". The "profligacy" shown by most animals may be explained if we allow that as many as 1 out of every 5 attempts at crossing over "go wrong". Animals without crossovers (for example the fruit-fly *Drosophila* and the bee) only produce tens of sperms for each fertilization. On the other hand in man, with 52 crossovers involved in the production of each spermatozoan, as few as 0.0009 per cent of the sperms produced may have got *all* the crossovers "right"; in 350,000,000 sperms, therefore, statistics suggest that there will only be about 3,000 with a full complement of successful crossovers. In other words, the huge numbers of sperms produced by man, and most animals, could be the result of an attempt to ensure a small population of "complete" gametes; the so-called "profligacy of nature" is the price we and other animals pay for the generation of genetic variety which is created as a result of a lot of crossovers. This recombination-produced variety can be exposed as the Mendelian ratios in genetical experiments.

Once one accepts the possibility that only a minute proportion of the sperms in an ejaculate may have got spermatogenesis completely "right", one has to consider the possibility that only *these* sperms may be capable of bringing about successful development after fusion with the egg. If this assumption is correct one has to consider the *probability* that some form of screening system will have evolved to minimize the chances of "wrong" sperms penetrating the egg. It is a well-recognized fact that in most animals, whether with internal or external fertilization, the vast majority of sperms are destroyed shortly after being deposited by the male. In animals with external fertilization, this is probably caused by the various chemicals released with or by the egg. In animals with internal fertilization, wholesale destruction is meted out to sperm by apparently normal processes within the female genital tract; in mammals, for example, the great majority of sperms are coated with antibodies, which permits white blood cells to phagocytose them. There is increasing evidence in some animals (from bedbugs and moths to guppies and rabbits) of a tiny proportion of sperms which do not attract this sort of attention within or outside the female's reproductive system; in mammals this is probably because these sperms show different surface properties from the majority and do not get coated with antibodies. It is tempting to suggest that this tiny population of sperms, which because they survive are necessarily much more fertile than the majority, are the "supersperms" with a full complement of correct crossover.

If one accepts the possibility that the female is screening the male's gametes to prevent those with these particular problems (but not, of course, with *any* genetic problems!) getting too close to the egg, the question arises as to how screening of a similar sort might be occurring with regard to the female's own gametes. Crossovers are also involved in egg production and if the mechanism of crossing over is inherently faulty we must assume that a large proportion of eggs will also be "wrong". In *r*-adapted animals, where large numbers of eggs are produced, it is possible that selection of viable eggs is left to the environment in which the zygotes develop. However, in *K*-adapted animals, where there is a lot of parental investment in the few eggs produced, it seems likely that evolution will have ensured a high success rate for them. In *K*-adapted animals, the eggs actually released from the ovary are never more than a tiny proportion of the oogonia originally present when the ovary is first formed. In humans, for example, about 400 eggs may be ovulated during a woman's reproductive life, out of a total of 5,000,000 oocytes present in the ovaries 4 months before birth. The eggs which are not ovulated are reabsorbed in the ovary **(oocyte atresia)** and *may* represent the female's screening of her own gametes, comparable to the screening of the sperms.

FERTILIZATION

Fertilization always occurs in a liquid medium and the sperms of externally fertilizing forms are often trapped by substances released with or by the egg. In some cases, these substances agglutinate sperms (make them stick together) and so temporarily prevent them from moving away again. Other substances secreted by eggs increase the motility of the sperms. These processes are said to assist in ensuing fertilization as sperms are temporarily trapped in the vicinity of the egg. In internally fertilizing forms, especially mammals, but also in fish, bedbugs and leeches, nearly all of the sperms are immobilized or agglutinated and then phagocytosed by special cells. The survivors, which probably have different surface properties from the rest, go on to fertilize (but see previous section).

The role of the sperm in the determination of bilateral symmetry (which part of the egg will be the front of the animal) has already been mentioned. In some cases (as well as the obvious genetic function of the fusion of the nuclei of the egg and sperm) this appears to be the main function and such eggs (e.g. sea-urchin, frog) will develop satisfactorily if pricked with a fine needle. This **artificial fertilization** works especially well if the medium contains detergents, proteases, lipases or any of a great variety of such biologically active substances.

The amoeboid sperm of nematodes enters the egg complete (Plate IId), as does the flagellate sperm of mammals. Other flagellate sperms cast off their tails before the head and middle piece proceed through the vitelline membrane (which has been "softened" by the substances contained in the acrosome) into the egg cytoplasm; Fig. 6 and Plate Ia–e. Parts of the middle piece and perhaps of the head cytoplasm then contribute to the **sperm aster** (Plate IIc), a centriolelike body which escorts the sperm nucleus to the egg nucleus and contributes to the spindle whish may be formed during fusion of the nuclei; it usually remains for the first division.

At the point of sperm entry many of the **cortical granules** begin to break down and contribute their contents to the fluid-filled cavity which

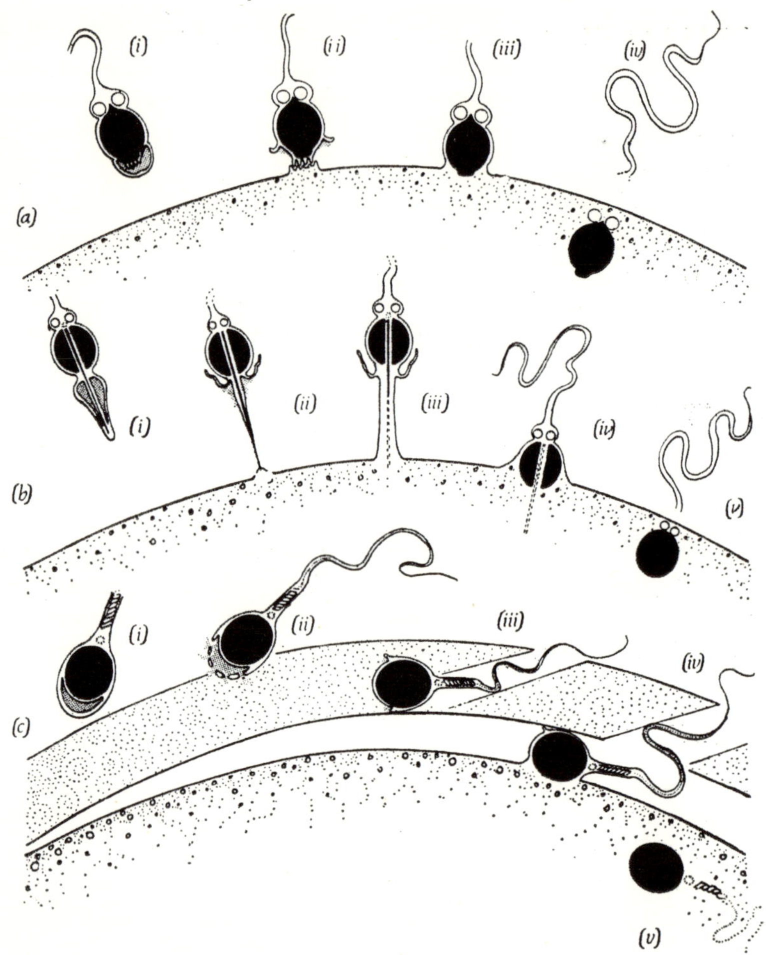

FIG. 6. *The process of fertilization. Note the dramatic acrosome reaction in the sperms of invertebrates* ((*a*) *a marine annelid,* (*b*) *a mussel*) *and the very different process in mammals* (*c*).

comes to separate the cytoplasm of the egg from the vitelline membrane (the **cortical reaction**). This membrane, due to a reaction with the contents of the cortical granules becomes a **fertilization membrane;** the reaction with the cortical granules probably involves a tanning of the protein of the vitelline membrane. The newly formed fertilization membrane usually moves clear of all the cytoplasm (Plate IV) and assists in the formation of a barrier to the penetration of further sperms. **Polyspermy** usually results from so high a concentration of sperms that several attach before this process is complete; however, it takes so long in large eggs that polyspermy is often the rule here (e.g. birds, fishes, cephalopods) and mechanisms inside the egg must ensure that only one male **pro-nucleus** unites with the egg nuclear material. In forms like the frog, which have a relatively large egg, polyspermy is effectively prevented in nature by the barrier presented by the albumen; sperms must struggle through this so that the chances of two arriving together are considerably reduced. This may also select sperms. In mammals, the **zona pellucida,** which represents the vitelline membrane, allows the passage of a few sperms prior to showing a **zona reaction** which prevents further sperms entering. In mouse and man, 1 or 2 sperms usually penetrate the zona, but more than 50 are found inside the zona of each rabbit egg.

It is extremely probable that the cytoplasm of the nematode egg is already completely organized, even before the gamete nuclei fuse, and so development is in every way **determined** from this time on; that is to say, each part of the egg can only produce one part of the adult. Even the number of cells which constitute each of the adult organs is constant for every species **(eutely)** and cell division does not occur after hatching, except in ovary and testis. Cases have been reported where isolated parts of the nematode egg, provided with a nucleus, have developed into perfect parts of the adult or larva, as if the rest had been present. Even mutations which cause developmental abnormality usually restrict their effects to clearly defined parts of the embryo, leaving all the rest developing normally.

Most eggs, however, do not show any such complete determination before fertilization; many parts of the cytoplasm can become any part of the embryo. In these, some cytoplasmic streaming is correlated with the fusion movements of the gamete nuclei and this may often be followed microscopically by the movement of pigment and yolk granules. These cytoplasmic movements, which are initiated by the events following the entry of the sperm or simply by ovulation, have been shown to commence in many eggs as a movement of yolky (usually vegetal) cytoplasm to an area opposite the path of sperm entry; as a result, there is movement of subcortical cytoplasm towards the animal pole and also towards the deeper parts of the egg (Fig. 7 and Plate IV). Naturally, this results in the

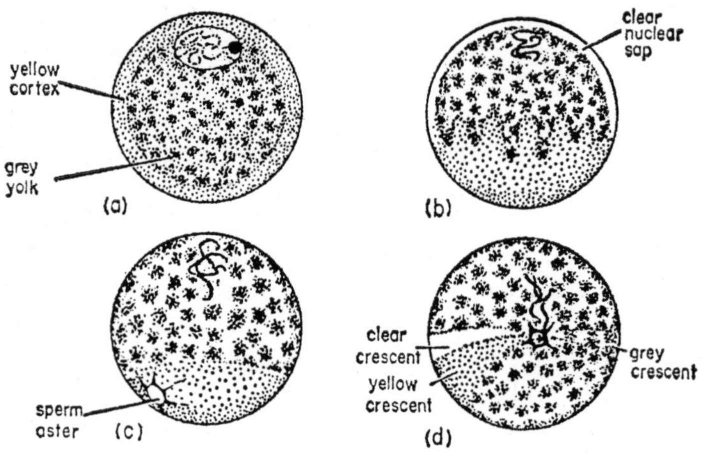

FIG. 7. *The egg of* Styela, *before* (*a*) *and during fertilization* (*b*), (*c*), (*d*). *The movement of cytoplasmic areas may be clearly seen in this egg.*

appearance of many new areas in the egg and *is the first stage in the epigenetic acquisition of complexity by the animal.* It has been called **ooplasmic segregation,** as the various substances (ooplasms) characteristic of the various parts of the adult were thought to be "sorted out" at this time.

CLEAVAGE AND ORGANIZATION OF THE EGG

All Metazoa consist of large numbers of cells, and cell divisions must therefore occur during their embryology. It should be noted that in almost all cases these **cleavage divisions** are *not* comparable with the multiplication of protozoans or with cell multiplication in the adult. The

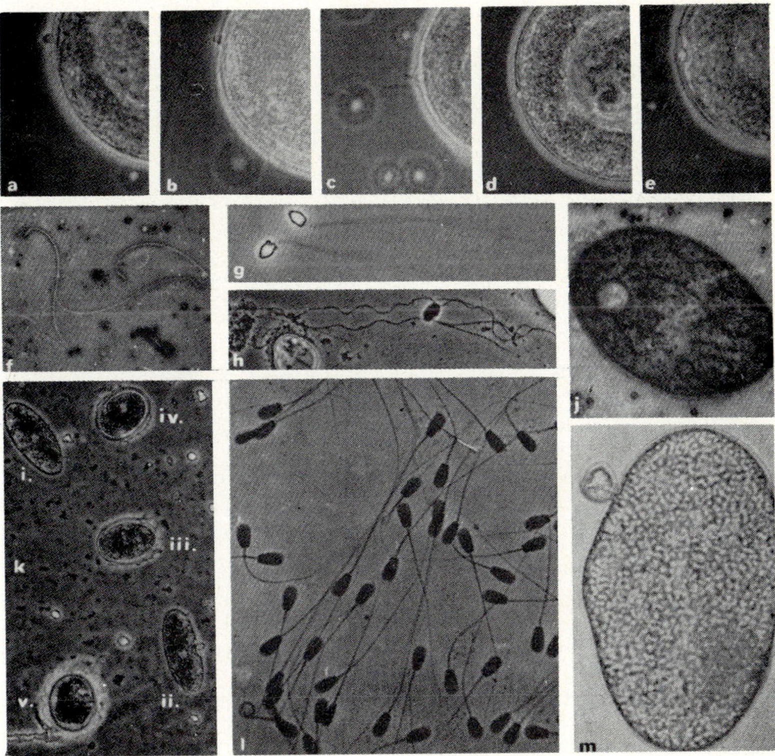

PLATE I. (*a-e*) *A sperm entering an egg of* Pomatoceros. *The tail is shed in* c. (*f*) *Rat sperms in Fallopian tube fluid. Note the "arching" movement and the hooked head.* (*g*) *Sperms of* Pomatoceros. (*h*) *Sperms of* Fasciola. (*j*) *Egg of* Fasciola; *the vitelline cells look dark and the actual egg cell light, and sperms are present too.* (*k*) *Stages in fertilization of* Ascaris *eggs.* (*l*) *Bull sperm; note the abnormal (curly) tail at bottom left, and the edge view of the head near the top.* (*m*) *Amoeboid sperm of* Ascaris *stuck to outside of egg (this egg may in fact already have been fertilized at bottom right).*

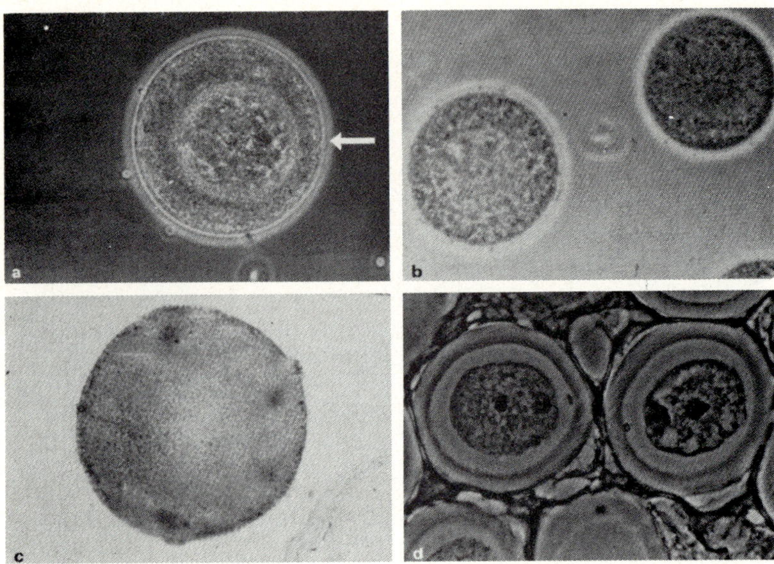

PLATE II. (*a*) *Supernumary sperms attached to the membrane of a* Pomatoceros *egg; one sperm head has already entered at the position of the arrow (note the wavy, or ridged, appearance of the membrane at this point, and compare with Plate Ie). The nuclear membrane is disappearing and the fertilization membrane is lifting very slightly (it never does lift far in this species, in normal sea water).* (*b*) *Two fixed, stained and cleared eggs of* Pomatoceros. *The one at the left is comparable to* (*a*), *but the other has proceeded further and the chromosomes (dark dots) may be seen lining up on the polar body spindle in the centre of the egg. The position of extrusion of the polar body will show which is the animal pole.* (*c*) *Section of a starfish egg into which at least 60 sperms have penetrated, although monospermy is the rule in this species; this egg would not have continued normal development. At least three sperm asters can be clearly seen in this section; that at the top of the section shows the sperm head clearly.* (*d*) *Part of a section of the oviduct of* Ascaris *at the level where eggs have just been fertilized. Note the amoeboid sperm proceeding into the cytoplasm whole, and the thick fertilization membrane. The dark mass in the centre of each egg is its mass of chromosomes remaining from the polar body divisions, which occur before the sperm enters, unlike* Pomatoceros *and most other animals.*

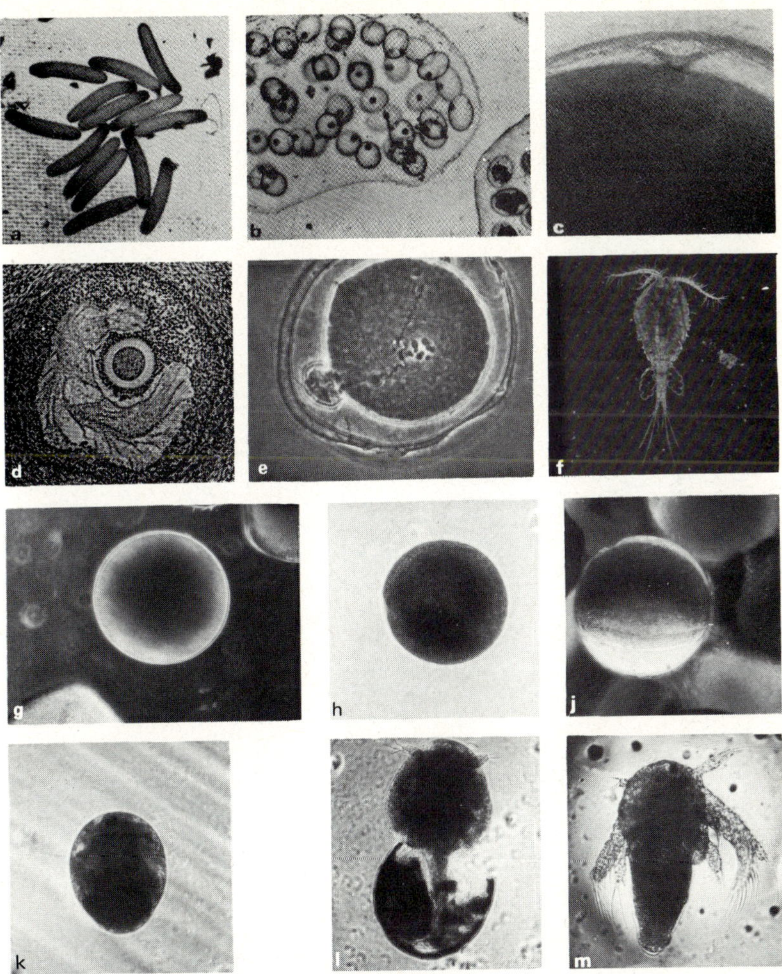

PLATE III. (*a*) *Eggs of locust removed from sand, a little larger than actual size.* (*b*) *Parts of two egg masses of* Limnaea, *a water snail. In the mass on the left are late cleavage stages, and on the right are little snails.* (*c*) *The cytoplasmic* (*animal*) *pole of a fish egg* (Hemichromis) *to show the funnel in the chorion by which the sperms enter, the micropyle.* (*d*) *Late stage of development of a mouse egg in the ovary.* (*e*) *A mouse egg, fertilized in culture. The two pronuclei and a polar body can be seen clearly.* (*f*) *A female* Cyclops *with egg sacs.* (*g-m*) *Development in* Artemia. (*g*) *Egg in diapause, as bought from a pet-shop. Note thick shell.* (*h*) *Development has begun and cleavage planes can be made out* (*with some difficulty*). (*j*) *Later cleavage. The dark area at one end of the egg is the future optic region: the future cephalic segment fills half the egg* (*see also Fig.* 22 (*b*)). (*k*) *The eye has condensed to a single black spot and 3 segments are clearly visible.* (*l*) *The nauplius hatching.* (*m*) *A nauplius larva* (*see also Fig.* 22 (*c*)).

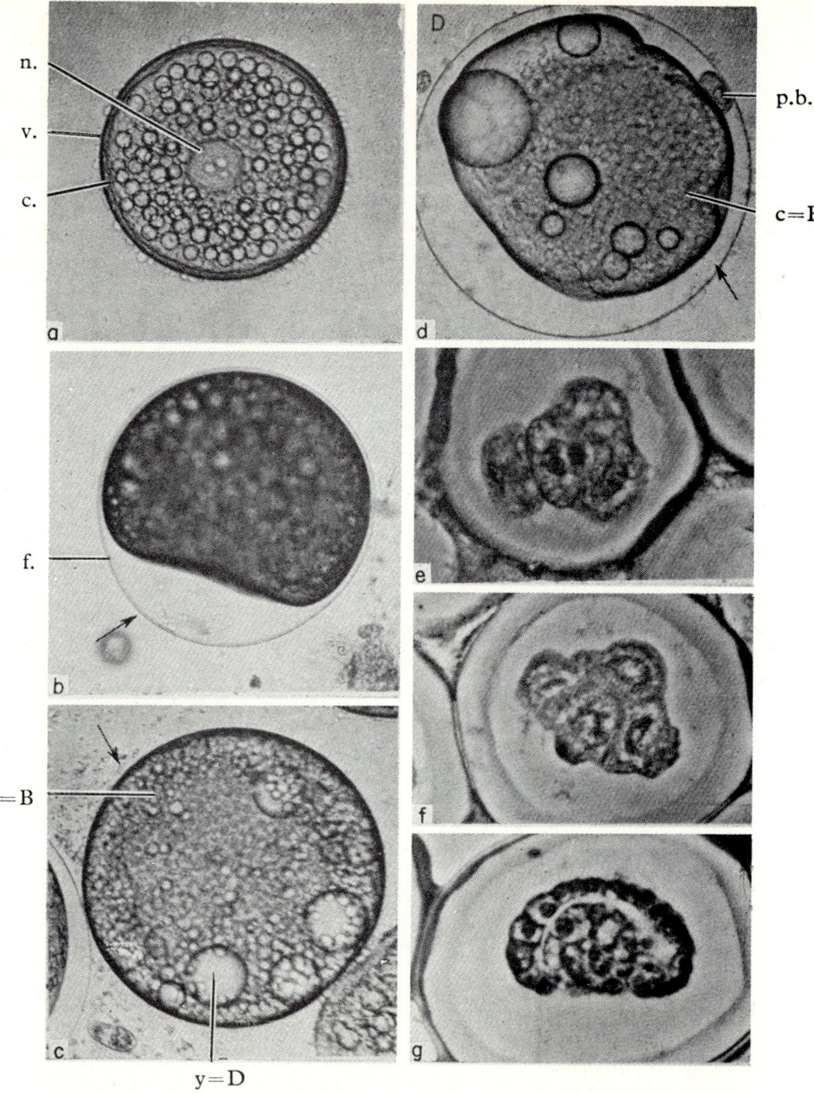

PLATE IV. (*a-d*) *Egg of* Nereis (*a polychaet worm*) × *600.* (*a*) *Before fertilization; note uniform distribution of yolk droplets in cytoplasm, though not in cortex.* (*b*) *and* (*c*) *Just after fertilization; the arrow marks the point of sperm entry. In* (*b*) *the fertilization membrane is lifting. In* (*c*) *the fertilization membrane has not yet lifted but yolk is aggregating opposite the point of sperm entry.* (*d*) *Egg just prior to first cleavage.* (*e-g*) *Sections of the early embryo of* Ascaris × *800.* (*e*) *T-shaped 4-cell stage; note the thick shell.* (*f*) *Rhomboid 4-cell stage.* (*g*) *"Gastrula".* c = *cortex;* f = *fertilization membrane;* n = *nucleus;* p.b. = *polar body;* B,D = *future B and D quadrants, full of cytoplasm and yolk respectively.*

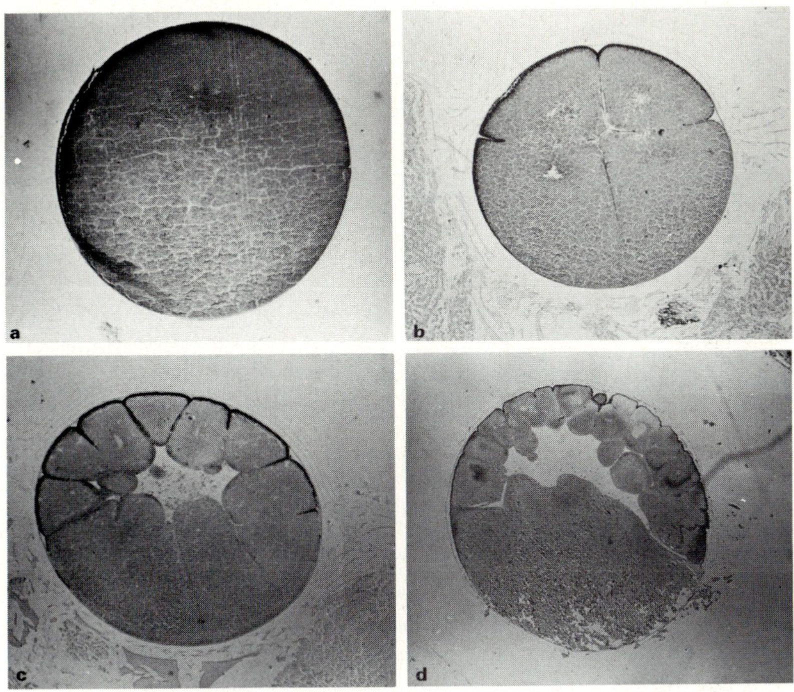

PLATE V. (*a*) *Vertical section of a just fertilized frog egg. Note the grey crescent at bottom left and the two nuclei upper centre (this may be the two gamete nuclei or the first division spindle); note also the pigmented cortex of the animal hemisphere (damaged during processing at left).* (*b*) *Vertical section of eight-cell frog egg. Note that the third cleavage plane lies above the equator.* (*c*) *Vertical section of early blastula of frog; note blastocoel and different size of cells.* (*d*) *Vertical section of lightly centrifuged frog egg. This treatment causes the heavier yolk to displace almost all of the vegetal cytoplasm; cleavage then only occurs in the animal hemisphere, as in telolecithal eggs.*

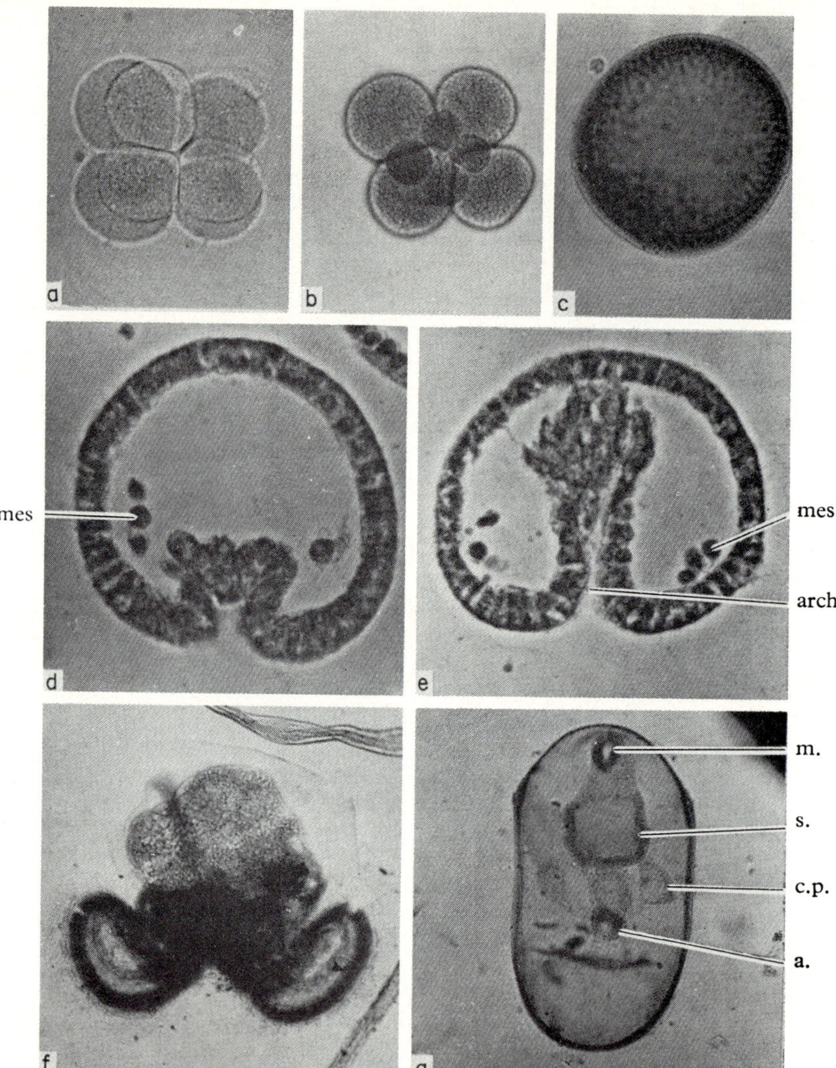

PLATE VI. (*a*) *An 8 cell stage of the sea-urchin* (Echinus) *from the animal pole. The two tiers of cells lie one above the other, i.e. this is radial cleavage* × *200.* (*b*) *Eight cell stage of the mollusc* Archidoris. *The micromeres alternate with the macromeres.* (*Mr. O. G. Harry prepared this beautiful specimen and I wish to express my gratitude. JC*). (*c*) *The late blastula of* Echinus. *Compare Fig. 24* (*a*) × *200.* (*d*), (*e*) *Two later stages in section, of gastrulation in* Echinus. *Compare Fig. 24* (*c*), (*d*), (*e*). (*f*) *The veliger larva of* Archidoris (*compare Fig. 20*). (*g*) *A young echinoderm larva, from the ventral side. Mouth* m. *anus* a. *and stomach* s. *are marked, also the coelomic connection with the "outside world", the coelomic pore,* c.p. × *50.*

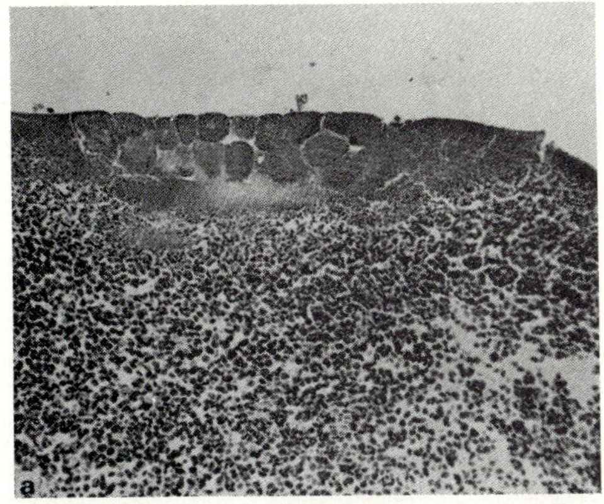

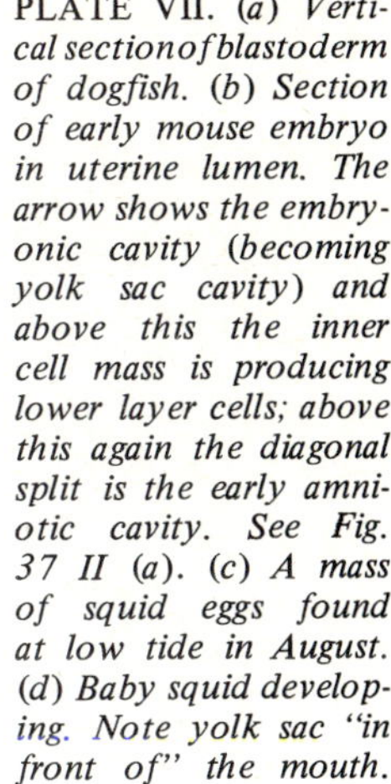
PLATE VII. (*a*) *Vertical section of blastoderm of dogfish.* (*b*) *Section of early mouse embryo in uterine lumen. The arrow shows the embryonic cavity (becoming yolk sac cavity) and above this the inner cell mass is producing lower layer cells; above this again the diagonal split is the early amniotic cavity. See Fig. 37 II* (*a*). (*c*) *A mass of squid eggs found at low tide in August.* (*d*) *Baby squid developing. Note yolk sac "in front of" the mouth.*

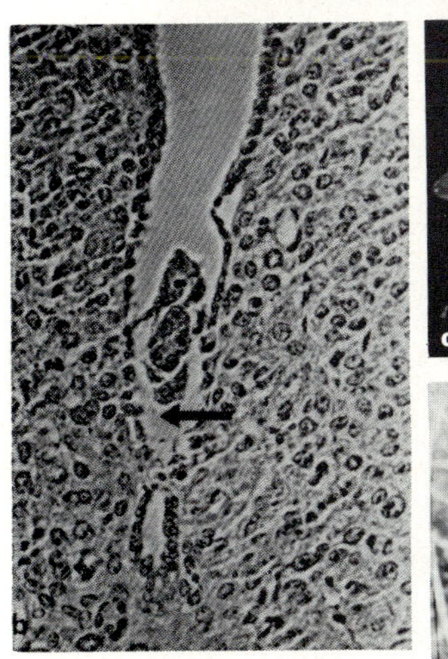

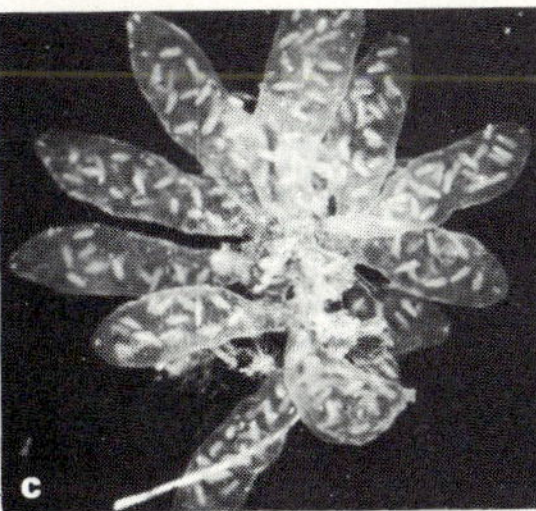

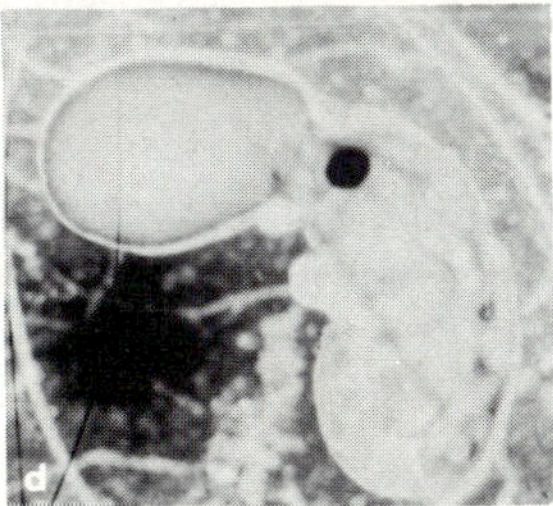

PLATE VIII. Development of Chick. (*a*) *Transverse section of primitive streak, compare Fig. 30* (*a*) *iii.* p.g. = *primitive groove,* spl. = *splanchnopleure,* som. = *somatopleure,* hyp. = *hypoblast.* (*b*) *A 30-hour chick (or slightly older). Primitive streak* ps., *anterior intestinal portal* AIP, *head fold and somites can be seen cf. Fig. 30* (*c*) *i. The neural folds* n.g. *are not closed posteriorly.* (*c*) *A 36-hour chick. The heart is S-shape, more somites have appeared, and the primitive streak is almost gone. Compare Figs. 30, which shows younger, and 31, showing older, stages.* (*d*) *Two longitudinal cuts were made in a chick blastoderm. The intucking from the primitive streak was split into three paths each of which formed a separate chick trunk and head. Two can be clearly seen, the other lies under these.* (*e*) *A chick at 72 hours. Note eyes, cranial flexure, tail fold, and especially the veil of amnion forming a tunnel over the head end* (*a*). *Compare Figs. 31 and 34.*

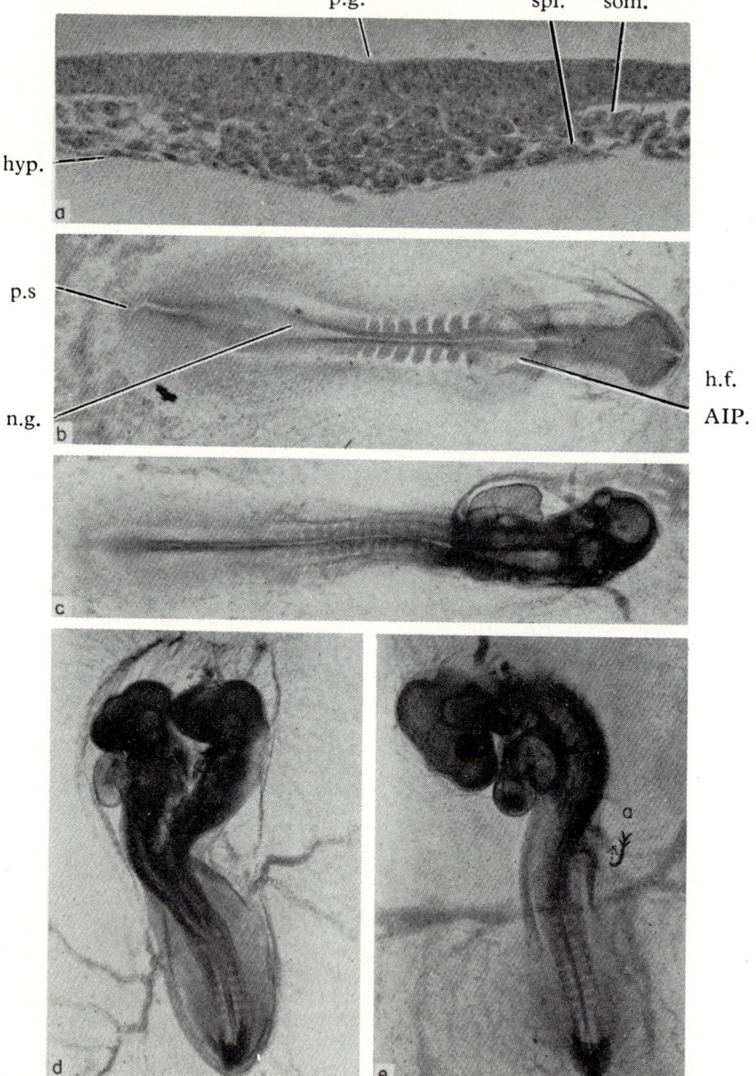

p.g.
spl.
som.
hyp.
a
p.s
n.g.
b
h.f.
AIP.
c
d
e
a

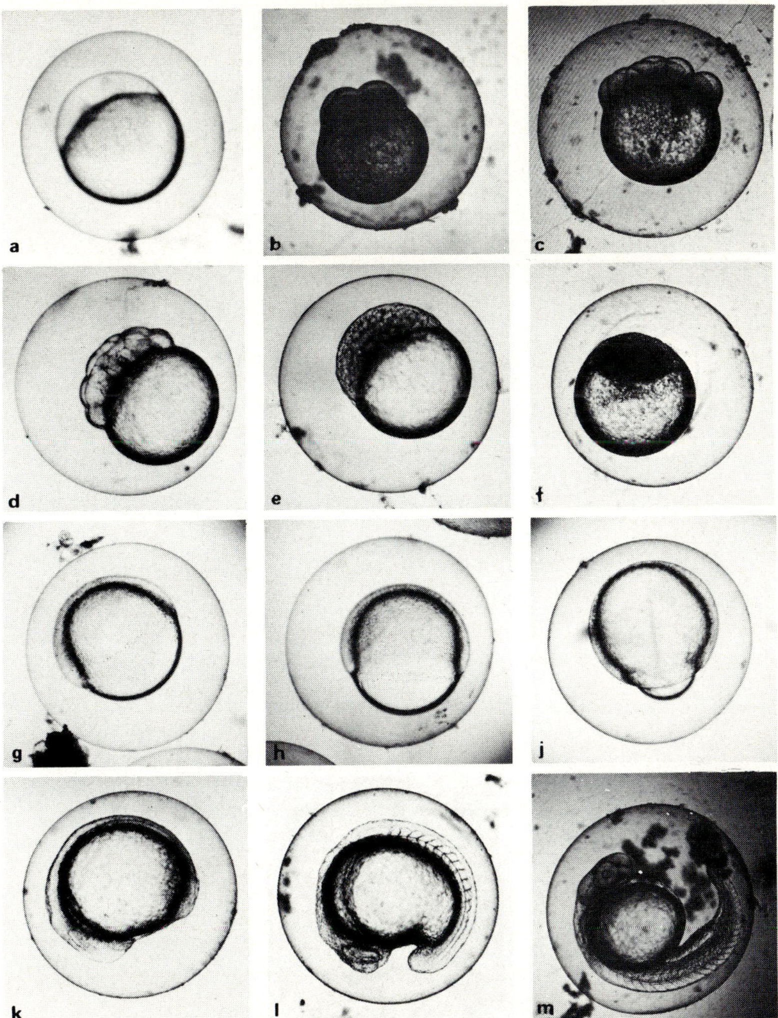

PLATE IX. *Development of zebra-fish* Brachydanio rerio. (*a*) *15 minutes after laying: the fertilization membrane is fully elevated.* (*b*) *60 minutes: 2–4 cells.* (*c*) *90 minutes: 16 cells.* (*d*) *130 minutes: about 64 cells present and periblast nuclei visible under blastodisc.* (*e*) *4 hours: blastula sitting on yolk.* (*f*) *6½ hours: the mass of cells has flattened and is spreading over yolk.* (*g*) *8½ hours: the spreading continues.* (*h*) *8½ hours: another egg from a different angle.* (*j*) *10 hours: only the "yolk plug" protrudes.* (*k*) *14 hours: embryonic axis now clearly visible (see Fig. 27(c)).* (*l*) *21 hours: eye cups, somites, heart and tail now obvious and the little fish begins to wriggle.* (*m*) *48 hours: baby fish ready to hatch: note how much of the yolk has now been used.*

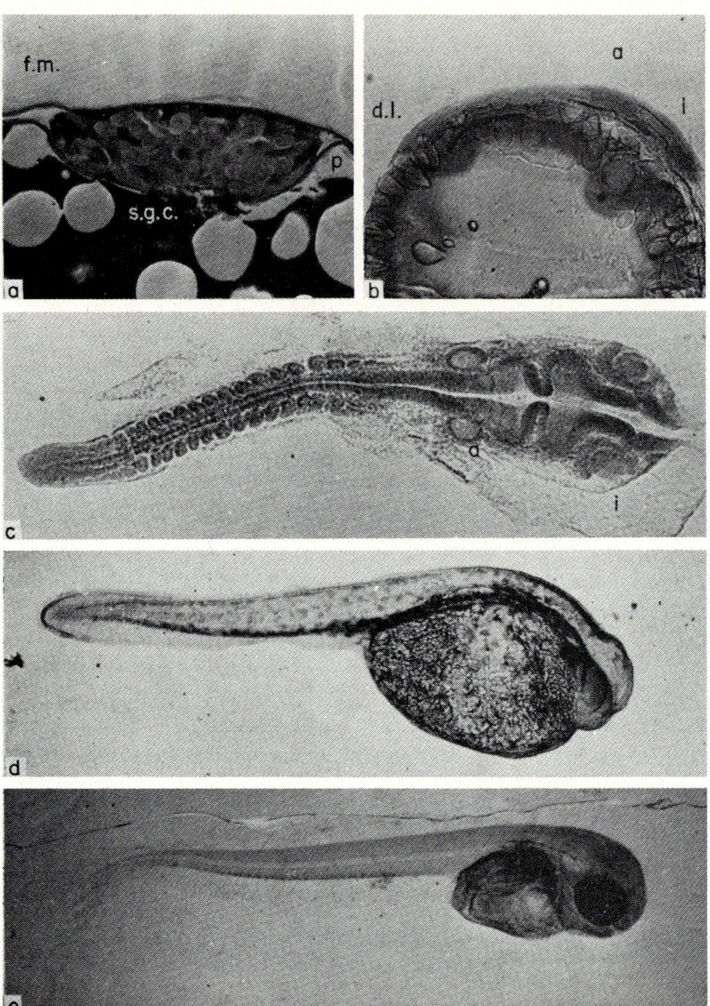

PLATE X. Development of Fishes. (*a*) *Section of the cleaved blastoderm of a guppy. The fertilization membrane* f.m., *sub-germinal cavity* s.g.c., *are marked, as is the layer of periblast* p, *delaminating from the underside of the blastoderm. The spaces in the sectioned yolk contained oil droplets.* (*b*) *A thick slice of an egg of the Platy* (Platypoecilus maculatus) *showing a young fish.* d.l. *marks the advancing edge of the blastoderm over the yolk,* not *forming notochord etc. The eyes* i, *and the ears* a, *are just appearing. Compare Fig. 27* (*b*), (*c*). (*c*) *A slightly later stage removed from its yolk and flattened. Note the subdivisions of the brain; eyes* i *and ears* a *have appeared. Compare this with Plate VIII* (*b*). (*d*) *A 3-day old Paradise fish* (Macropodus opercularis) *i.e. 1 day after hatching. Note the large yolk sac.* (*e*) *Paradise fish, 3 days after hatching. Eyes and ears are now well-developed.*

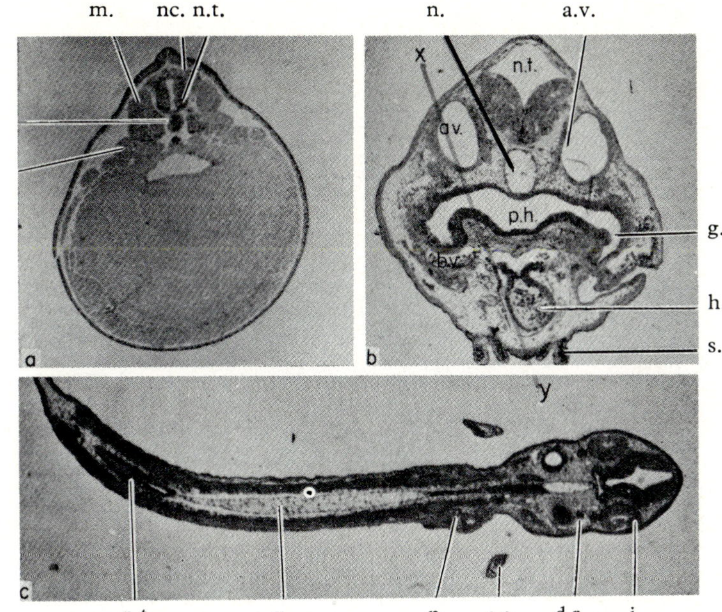

PLATE XI. Organ Formation in Amphibia. (*a*) *Transverse section of late neurula of axolotl. Neural tube* n.t., *notochord* n., *myotome* m., *nephrotome* k., *and neural crest* n.c. *are marked.* (*b*) *Transverse section of young tadpole through gills and heart. Pharynx* ph., *auditory vesicles* av., *neural tube* n.t., *notochord* n., *heart* h., *and a branchial vessel* b.v. *passing on the left side* between *gill pouches, are marked. On the right a gill pouch* g. *has been included. The sucker* s. *is also sectioned.* (*c*) *High* (*i.e. dorsal*) *horizontal longitudinal section of similar tadpole to* (*b*). *Eyes* i., *external gills* e.g., *dorsal roots of cranial nerves* d.r., *and pronephros* p. *in cardinal*

sinus are labelled. (*d*) *A lower* (*i.e. more ventral*) *section of the same tadpole. Nasal placodes* na., *pituitary gland* pit., *pharynx* ph. *with gill pouches, oesophagus* oe., *are labelled.* (*e*) *An oblique longitudinal section of a similar tadpole, considerably to the left of the midline, so spinal cord and brain do not show. The line XY on the transverse section in* (*b*) *shows the position of this section.*

Marked are: auditory vesicle a.v. *with vertical semicircular canal pinching dorsally and a dorsal root* (*nerve VIII*) *pressing on the anterior aspect; heart, very folded* h., *liver* l. *which has invaded the transverse septum along the ventral mesentery, and broken the old vitelline vein into an hepatic portal vein* h.p.v. *and an hepatic vein; the gut* g. *is still very yolky but the coelom* c. *is capacious and the dorsal mesentery can be clearly seen; a Wolfian duct* W.d. *passes in and out of the section, in the dorsal coelomic wall.*

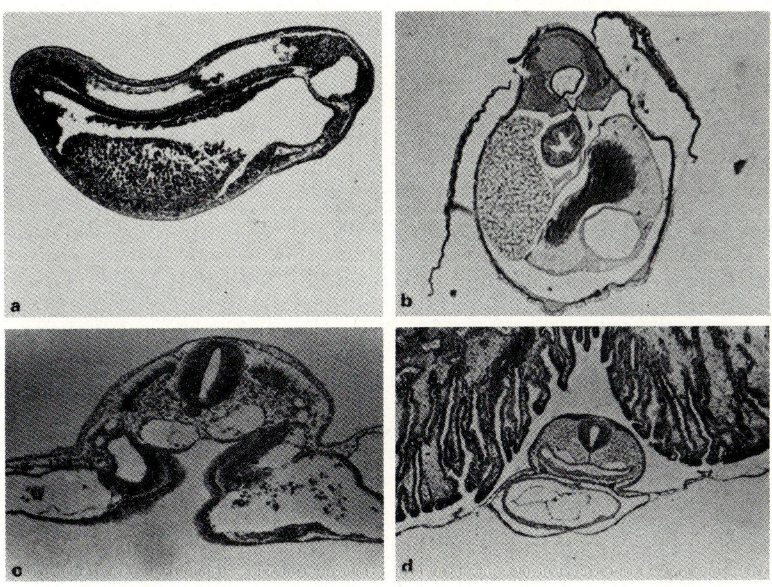

PLATE XII. (*a*) *Vertical longitudinal section of frog neurula (compare Figs 9 and 14*). (*b*) *Transverse section of newly-hatched (unidentified marine) fish; note mass of yolk in peritoneal cavity, not in gut.* (*c*) *Transverse section of chick embryo just posterior to A.I.P. Identify nerve cord, notochord, somites (on either side of nerve cord), gut (with no floor), paired dorsal aortae, nephric folds (on either side of dorsal region of gut). The extra-embryonic coelom here is filled with the enormous vitelline veins, and corpuscles can be seen.* (*d*) *Transverse section of rabbit embryo in uterus, just anterior to the A.I.P. Identify nerve cord, tiny notochord, gut (= pharynx)* with *a floor at this level, paired aortae above the gut, and the enormous heart in pericardium. Note also uterine villi and trophoblast wall.*

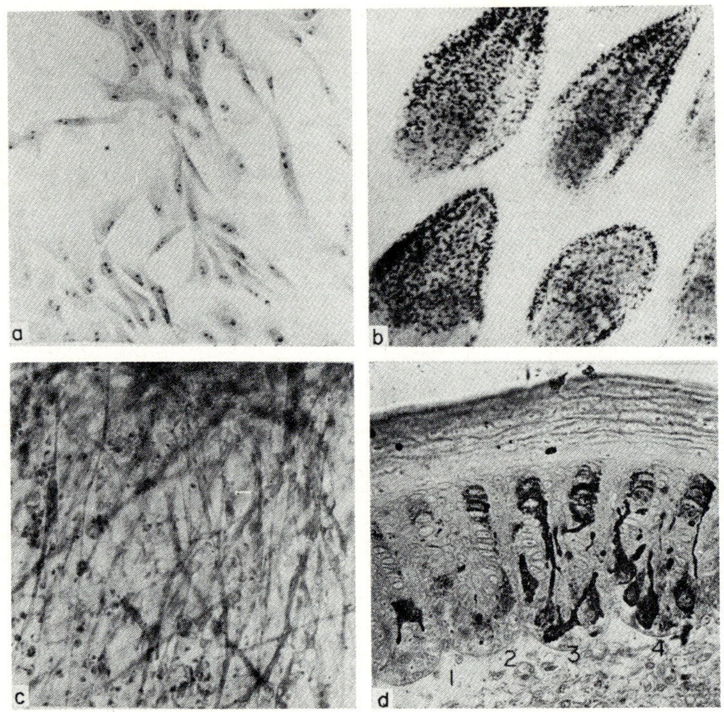

PLATE XIII. (*a*) *Cells of chick heart in "unorganized" tissue culture. Note that the "amoeboid" shape is quite different from the normal.* (*b*) *An "organized" culture of chick skin in which feather germs have developed and are pigmenting normally.* (*c*) *Nerve cells and other neural crest cells in a tissue culture of chick neural crest.* (*d*) *A section of part of a developing Brown Leghorn neck feather. Melanocytes can be observed in barb ridges 3, 4, with their long processes packing melanin pigment into epidermal cells. In barb ridges 1, 2, three similar but unpigmented cells are just visible. These may be melanoblasts. Melanocytes with yellow pigment can be observed in barb ridge 2.*

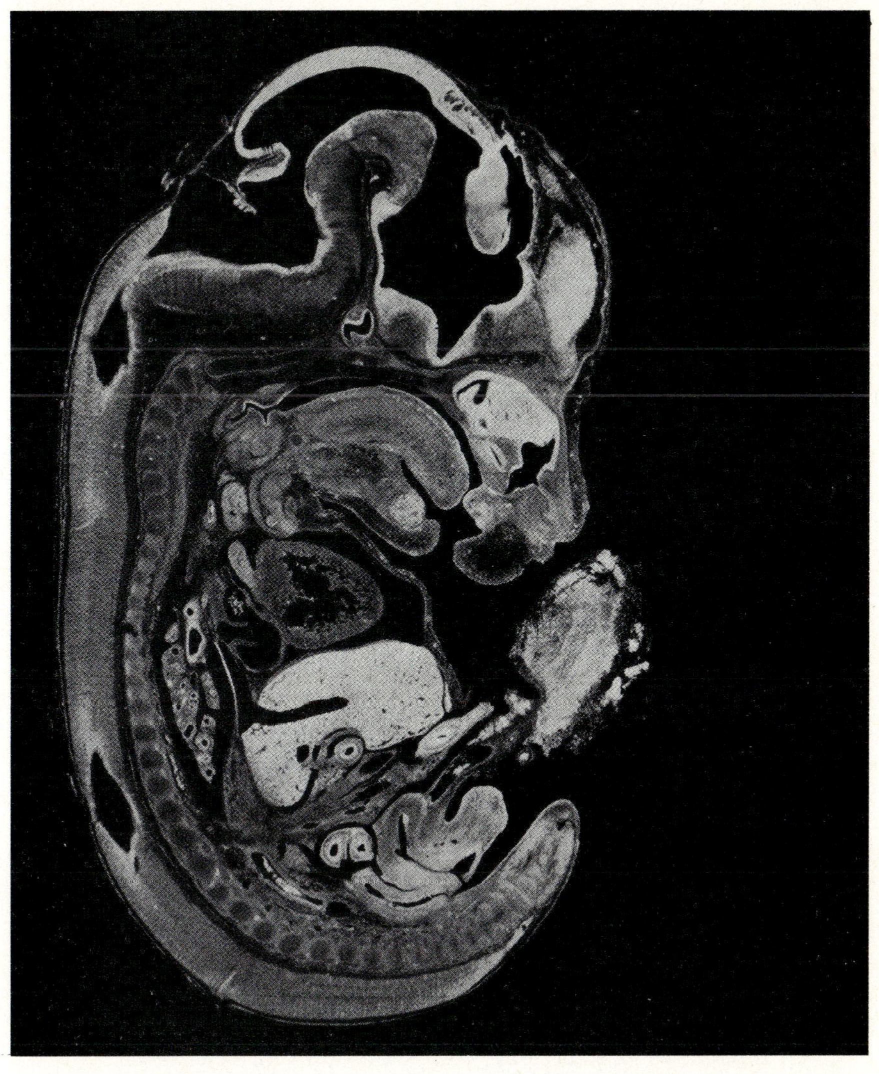

PLATE XIV. *Longitudinal median section of 19-day mouse embryo. Refer to outline print for labelling. An exercise in understanding of this section may be devised by covering the caption.*

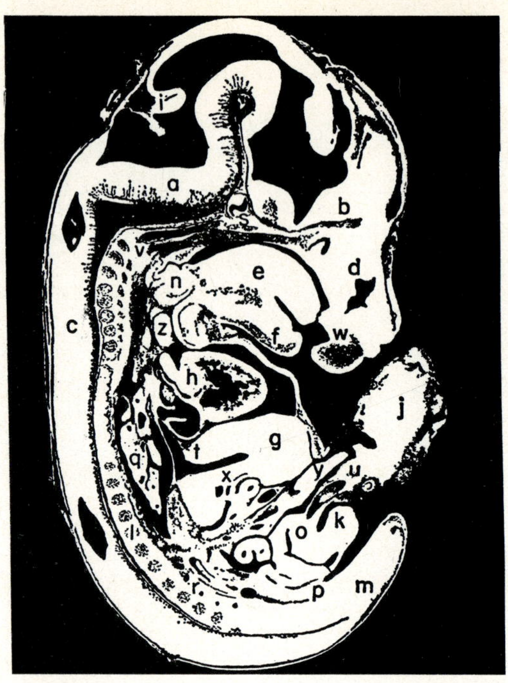

Key to PLATE XIV.

a. *Floor of hind-brain (medulla oblongata).* b. *Anterior wall of fore-brain.* c. *Spinal cord (with two holes where the section passes into the cavity).* d. *Nasal epithelium covering the turbinals.* e. *Tongue.* f. *Lower jaw.* g. *Liver, whose front end is closely applied to the transverse septum.* h. *Heart; the spongy wall of the ventricle, and the atrium with its valves, can be clearly seen.* i. *Choroid plexus.* j. *Umbilical cord, passing out of the plane of section.* k. *Primary genital papilla, which will become either the penis or clitoris. The urethra may be seen opening at its base.* l. *Vertebral column replacing the notochord.* m. *Tail.* n. *Position of the thyroid gland, and of the opening into the trachea.* o. *Front wall of the bladder, whose cavity is seen to be continuous down to the urethra.* p. *The rectum opening just below the tail as the anus.* q. *Lung in pleural cavity, whose front wall represents the dorsal part of the old transverse septum. Its hind end is composed in part of Uskow's septum and in part of the fused mesogastria supporting the stomach and gut.* r. *Part of the posterior vena cava between the kidneys, which lie on either side of this median section.* s. *Pituitary gland.* t. *The label is just above the wide hepatic vein from the liver into the atrium of the heart. (In this section a fold of the wall seems to occlude the passage of blood. It did not of course block the entire vessel and it serves to mark the passage of this vessel through the transverse septum.)* u. *Umbilical vein (allantoic vein).* v. *Anterior end of notochord and "pre-chordal plate".* w. *Upper jaw; the incisor rudiment may be seen on close inspection.* x. *The bile duct and gall bladder (?) between the pylorus of the stomach and the liver.* y. *Yolk sac stalk in umbilicus – the A.I.P. and P.I.P. which have met.* z. *Thymus gland.*

mitotic apparatus and process is similar, but there are two very important differences. Firstly the daughter cells, especially in determinate eggs (see below), are often very different in their contents and in their potentiality. (In cell multiplication, on the other hand, the products are all similar.) The nuclei of cleaving eggs are identical in DNA content, except in a few cases. During the development of the nematode *Ascaris* the chromosome number is reduced in various kinds of cells, a phenomenon peculiar to this very determinate kind of development; it also occurs in a few insects, for example the gall midges. In these cases, the future germ cell nuclei, of course, retain their full complement of chromosomes.

Secondly there is almost always *no increase in material* during cleavage. The sum of the masses of the daughter cells is usually a little less than that of the parent cell, because of the utilization of some material for energy and the diffusion of waste products. Therefore, it is misleading to speak of the egg as "one cell" except in a very restrictive and formal sense. The egg cleaves into cells; it becomes cellular. It does *not* recapitulate the origin of Metazoa from Protozoa by changing from unicellular to multicellular. (Indeed, the theory of metazoan origins put forward by Hadzi also *splits* a ciliate into a cellular metazoan.) There are two important exceptions to this almost universal rule. The gastropod molluscs usually lay virtually alecithal eggs in capsules of nutritive albumen (in jelly, Plate IIIb) and the mass of each embryo increases very considerably during development. The other exception is where **viviparity** occurs; this is the retention by the parent of the eggs inside its body until a very late stage of development, usually until they are more or less capable of fending for themselves. The condition has been divided rather artificially into: **ovoviviparity,** where the egg is merely enclosed in the body of the parent till it hatches; and true viviparity, where it derives nutriment as well as oxygen from the parent's body. It is the latter case which is the exception to the above rule. It has been shown that in the elasmobranch (selachian) fishes various degrees of viviparity exist. Where the egg has much yolk, there is little or no gain in dry weight while sojourning in the mother; on the other hand, there are some forms with little yolk where the dry weight increases during development by four or five times. The mammal shows the logical end of this series: the alecithal egg is shed into the Fallopian tube and almost certainly uses the secretion here for nutrition, even before attaching to the uterine wall and becoming a parasite proper. During this time, a few species have been shown to increase slightly in dry weight; later, of course, the increase may be of the order of 15,000,000 times in the human, and 5,000,000,000 times in the blue whale (weight at birth/weight of early cleaved egg).

The cleavage of the egg into "cells", the **blastomeres,** must be so arranged that any organization which the egg has achieved during and after fertilization, but before cleavage commenced, is not disrupted. In the annelid worms and the molluscs, almost all of the interactions between parts of the egg occur before cleavage commences and in consequence, if the blastomeres are separated at the 4-cell* or 8-cell stage, they have already been determined in their fates and can only produce parts of animals. This kind of development is called **"determinate"** or **"mosaic"** and is also shown by many primitive chordates, especially the tunicates or sea-squirts (Fig. 26). Other animals, notably the echinoderms and vertebrates, delay most of these interactions between parts of the egg until cleavage has progressed far enough for movements and interactions again to become possible. (It is obviously difficult to do this kind of thing with 2, 4, 8 or even 16 blastomeres.) This mode of development is called **indeterminate** because single blastomeres taken from early cleavage stages can be made to give rise to whole embryos, naturally of small size.

Fertilization usually starts the processes both of development and cleavage, but the eggs of *Chaetopterus*, a polychaete worm, may be made to develop almost normally without any cytoplasmic cleavage. Cleavage is therefore regarded by many embryologists as a process parallel with, but not contributing to, the process of organization of the embryo. In the egg of the mouse, too, it has been shown that progressive cell changes, for example the compaction usual at 8-cells, occur at a particular age and are independent of divisions.

In determinate eggs, of course, the cleavage must so occur that the various already organized parts of the embryo come to lie in predictable

* We are forced to retain this usage because it is so well entrenched—it should really be 4-blastomeres of course!

positions; we might, therefore, expect complicated cleavage patterns to appear. The spiral cleavage of annelid worms and molluscs is a good example of this (p. 40) and, because the cells all come to lie in exact positions and are already determined, it is useful to label them (see Figs. 16 and 17). In indeterminate eggs, however, the individual cells have more or less equal potential for development and so labelling is not useful. Indeed, indeterminate development is usually easier to consider in terms only of the movement of material; the fact that this material consists of cells is incidental.

Before considering the development of specific animals a resumé must be given: the egg of the nematode is determined in its development and in the development of its parts even before fertilization is complete; many other invertebrates, for example worms and molluscs, organize their cytoplasm just after fertilization and their parts are determined in their fates after only one or two cell divisions; the blastomeres of other animals, on the other hand, notably echinoderms and vertebrates, each remain able to form a complete embryo for a considerable time after cleavage begins and many of the epigenetic events which pattern the embryo are delayed.

CLEAVAGE GEOMETRY

All of the Metazoa are thought to have arisen from a common ancestor, except probably the sponges and possibly the Aschelminthes (the nematode worms and their relatives). Therefore we may attempt evolutionary comparisons between their eggs and embryos at various stages of development. The earliest structure about which there is any controversy is the **blastocoele,** the cavity which arises within the group of blastomeres into which the egg has divided. It usually makes its first appearance at the 8-cell stage when, as can easily be verified by experiment, the more or less spherical blastomeres must have a space in the centre of the group. For purely geometric reasons, this space usually appears at the level of the third division (Fig. 5). Let us consider this for a moment. In very nearly all animal eggs (except the insects*), the first division is vertical, from animal to vegetal pole, and divides the future embryo into right and left halves (although it may not divide the whole egg). It will be seen that this must pass through 0° longitude (where the sperm entered) and 180° longitude, because the point of sperm entry defines (or is defined by) the plane of bilateral symmetry. Except in a few very yolky eggs, the second division is also vertical, but at right angles to the first (see Fig. 5).

The third division is usually a horizontal division, at right angles to the previous two and, in homolecithal eggs, is in the plane of the equator and the egg nuclei; 8 blastomeres of equal size result. In mesolecithal eggs, the third division is more animal (for example in the frog it is at about the Tropic of Cancer) and the 4 animal blastomeres are therefore smaller than the 4 vegetal ones.

In telolecithal eggs, the situation is complicated due to distortion of their geometry by the enormous masses of yolk. In fact, because the nucleus tends to be more animal in very yolky eggs and because the actual cytoplasmic volume (not counting the yolky volume) is divided fairly equally at each division, the third cleavage plane almost always passes through the position of the original zygote nucleus.

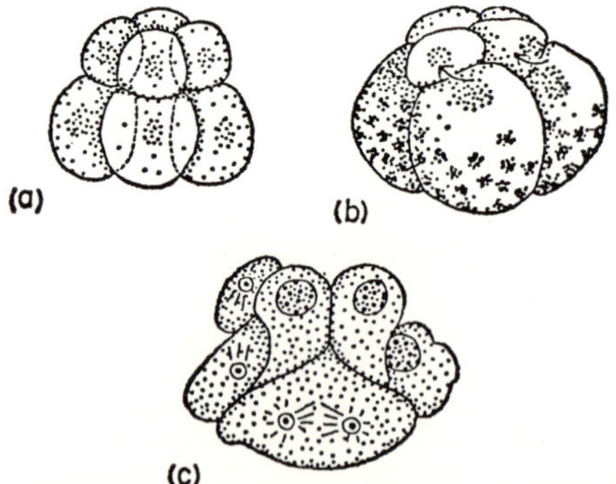

FIG. 8. *8-cell stages of (a) an echinoderm, with radial cleavage. (b) An annelid worm, with spiral cleavage. (c) A nematode, with aberrant cleavage, i.e. not comparable with other animals.*

* But the insects constitute 75% of the named species of animals!

At the 8-cell stage two families of cleavage types (compare Fig. 8) can be recognized. In **radial cleavage,** the four upper cells each lie directly above each of the four lower cells. In **spiral cleavage,** each "animal" (upper) blastomere lies between two "vegetal" (lower) blastomeres; that is to say, the mitoses which produced them were angled upwards. These angled divisions are alternately clockwise and anticlockwise (see Fig. 16). All spirally cleaving eggs are determinate and this form of cleavage appears to have been developed to enable higher precision in the distribution of the already organized cytoplasm. Some radially cleaving forms also have determinate eggs (for example *Branchiostoma*, tunicates, ctenophores and perhaps some coelenterates), while other radially cleaving forms are indeterminate (for example vertebrates, echinoderms).

As might be expected, eggs show many kinds of **aberrant cleavage** forms; some are obviously derived from radial cleavage types (for exam ple brittle stars) and others (oligochaets) from spiral cleavage types. The cleavage of nematodes, whose embryology is so very different from that of other animals, is impossible to consider in terms of radial or spiral. The development of Platyhelminthes is probably a derivation of spiral cleavage (indeed the polyclad flat-worms show almost typical spiral cleavage), but the development of the parasitic flatworms and tape-worms is complicated and very difficult to compare with other forms, perhaps partly because of the insertion of **"vitelline cells"** into the egg-shells with eggs and sperms.

The formation of the blastocoele of most invertebrates is relatively simple but, especially among the vertebrates, the complications arising from the extraordinary geometry of the telolecithal egg make the problem altogether more difficult. The disc of cytoplasm of these eggs always digests the yolk beneath it, so that in histological sections a space, the **subgerminal cavity,** appears, apparently where one might expect to find the blastocoele. Furthermore, the first few cleavages are usually all vertical and only later do horizontal cleavages appear; the plane of the third cleavage is thus of no use to us for comparison with other types. However, reference to Fig. 5 in which cleavages in fish, frog, and echinoderm are shown, may clarify the situation.

GASTRULATION

The group of cells enclosing the blastocoele has been called a **blastula** and the expression **gastrulation** has been used for the process by which this blastula is converted into the primitive body-form of the animal. In many ways the use of one word for diverse processes is misleading. All processes by which the spherical blastula is converted into a two-or-more-layered embryo have been called gastrulation. The mechanical folding-in of the gut of the already organized worm embryo is a very different process from the gastrulation of an indeterminate form like the frog, where a very complex system of **epigenetic relationships** and movements results in the production and initial differentiation of the embryonic organ systems. If one is forced to compare some process in determinate eggs with gastrulation in such indeterminate forms as the frog, one must surely choose the cytoplasmic movements following fertilization, at which time the epigenetic relationships between parts of the egg are occurring. Subsequently we shall use the word gastrulation to refer *only* to the movements *after* cleavage of the *indeterminate* egg which result in the organization of the embryo into organ systems whose cell potentialities have been restricted.

Let us consider in some detail the gastrulation of the sea urchin. Cleavage of the homolecithal egg produces a hollow ball of cells, the blastula. The movements of gastrulation convert this into a two-layered structure, which is ready to complicate into the three-layered structure of the larval sea urchin (the *Echinopluteus* larva). The mechanism of this transformation is as follows. Cells at the vegetal pole grow taller and the vegetal pole flattens; some of these cells get squeezed out from between their neighbours (Fig. 24) and move into the blastocoele (developing long processes, see p. 55). Then the cells which are left at the vegetal pole themselves put out long processes which reach right across the blastocoele towards the undersurfaces of the cells at the animal pole. Most of these processes anchor themselves onto the inside surface of the animal end of the blastula and then they contract. This operation pulls the vegetal pole into the blastocoele as a long, thin, hollow finger of tissue (see Fig. 24). Where this "finger" makes contact with the inner aspect of the blastula wall a pore breaks through which forms the mouth of the larva. The hollow finger is now the gut and at the original vegetal pole is the anus. The skin and the gut have thus been formed, but the internal organs of the animal have yet to appear. Skeleton comes from the wandering cells in the blastocoele (see p. 56), but muscles and larval kidneys are produced from the walls of vesicles which bud off from the side of the gut in the same way as the gut itself budded from the vegetal pole.

The result of the **morphogenetic movements** of gastrulation which we have just described has been to bring about a multilayered embryo from a simple, essentially single-layered, sac of cells, the blastula. In the sea urchin, this is achieved in two basic stages: firstly, the gut is pulled in (stage 1); latterly, the mesoderm-filling between the gut and the epidermis is produced (stage 2). In many animals, especially vertebrates, both gut (endoderm) and the middle (mesoderm) layers are produced as a result of just one complex folding operation occurring at gastrulation. This undoubtedly makes the process more difficult for the student to visualize and understand, but the end product of gastrulation is always the same, the production of a multilayered basic baby animal from a single-layered blastula. The routes from various kinds of egg to the basic baby animals is varied, between phyla of animals and even within one phylum. When we come to the vertebrates (p. 69) we will consider the details, and the significance, of these different routes.

All of these complex processes are determined by the maternal genome in the nucleus of the oocyte, and by other properties of the egg determined by the mother during development of the oocyte in the ovary. This a very general situation (p. 68) and a great variety of animal eggs have now been shown to have an architecture and a provision of m-RNA from the maternal genome which determine the complex events of gastrulation; the zygote nuclei up to this point have not started to release m-RNA into the cytoplasm and are merely "passengers" in the newly formed cells.

PRESUMPTIVE OR FATE MAPS

In the example of gastrulation above, the gut of the sea urchin larva has been pulled inside from some of the cells which were originally on the outside. Let us imagine that we have a film of the development of the sea urchin gut; if we coloured the gut green and then ran the film *backwards*, keeping the colour "attached" to the gut material, we would finally arrive back at the blastula with a green-coloured patch on its outside around the original vegetal pole. This coloured patch represents the **presumptive** gut. If we had allowed the embryo to develop further and coloured the skin blue and the kidneys red and carried out the same trick with the film, we would have found that the blastula had blue and red areas on its outside as well as green.

This device can be used with any blastula to indicate which areas of the surface cells will eventually form certain tissues and organs; a diagram of a blastula thus marked with colours or labels indicating future development is called a **fate map,** because it indicates the prospective fate of

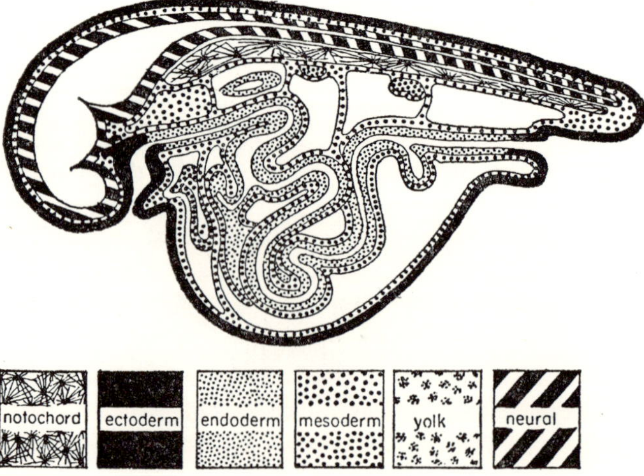

FIG. 9. *Diagrammatic vertical longitudinal section through a young vertebrate, to introduce the codes for fate maps.*

the blastula cells. In fact, we can also produce a fate map of the uncleaved egg; since the various areas of the egg do not change in relationship to each other during cleavage, the only difference between the fate map of an egg and a blastula is that the coloured areas on the blastula cover patches of cells, whereas the same areas on the egg are non-cellular.

In constructing fate maps, it is customary to use standardized colours to represent particular tissues. The system which is in most common use and which derives from the **germ layer theory** (see p. 37) is as follows: blue or black = ectoderm; red = mesoderm; green = endoderm; violet = nervous system; brown = notochord; yellow = yolk. Deep red or orange may be used to indicate systems within the mesoderm, for example somites, kidney, etc.

Since we cannot provide coloured diagrams in this book we have used various kinds of stippling in our diagrams to represent the tissues referred to above. Reference to Fig. 9 should make this clear.

(If difficulty is experienced in grasping the concept of the fate map a useful exercise is to deform rubber balloons, make large coloured patches on them—and then to return them to spherical shape.)

THE EARLY EMBRYOLOGY OF THE FROG

The early development of the frog will be discussed in order to give a basis for comparison with other species.

The egg of the frog, while in the ovary, comes to have the following structure. One pole, the future animal pole, has pigment granules just below the surface of the cytoplasm (in the cortex), while the future vegetal pole is packed with yolk granules and has no such pigment (Fig. 10). In the eggs of several kinds of frog there is a disposition to bilateral symmetry already possessed by the egg at this stage (see p. 12). This is manifested by a pigment pattern and by reluctance to develop in response to puncture, except in a certain area; in these forms the position of sperm entry is to a certain extent determined.

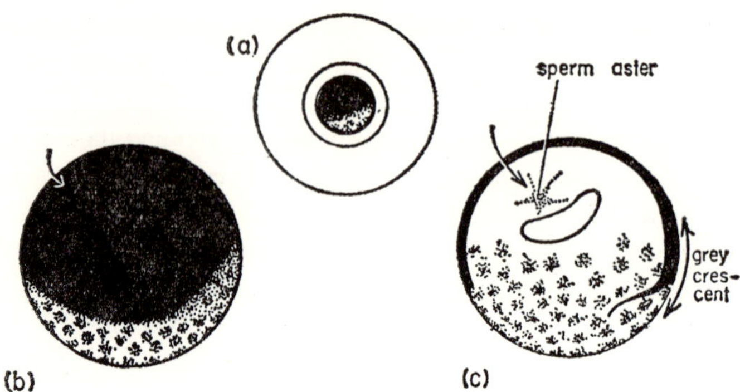

FIG. 10. *The frog's egg. (a) In its membranes. (b) After fertilization; the arrow indicates position of sperm entry and the grey crescent has appeared opposite to it. (c) Vertical section of (b).*

The egg is, of course, enclosed in a vitelline membrane, outside which is a dense layer of albumen which is probably bounded by another membrane. We may now name such membranes. The vitelline membrane, which is truly part of the egg-cell, is called **primary**. Membranes which are produced by ovarian tissues other than the egg itself are called **secondary;** examples are the jelly coat of sea urchin egg and the zona and cumulus of the mammal egg. (Some books call the embryonic membranes, see p. 75, secondary membranes but this practice should be discouraged.) Additions to the egg which lie outside the vitelline membrane, produced after fertilization, are called **tertiary** membranes, for example, the albumen of the frog and bird and the shell membranes and shell of the bird egg. Such membranes are produced by the lining of the oviduct or special glands of the female genital tract.

The egg of the frog is laid while the animals are pairing in water and the tertiary membranes begin to swell almost immediately. The water in which they are laid has a high concentration of sperm, many of which appear in the albumen and make their way to the egg surface. One (or, rarely, more) penetrates the vitelline membrane and enters the cytoplasm of the egg. Commencing at this point and spreading over the surface in all directions, the vitelline membrane transforms into the fertilization membrane and elevates from the surface of the egg. The egg is now free to rotate under the influence of gravity, so that the denser yolky vegetal pole comes to lie at the bottom, the pigmented animal pole lying uppermost (Fig. 10 (a)). The path of the sperm nucleus to the the egg nucleus usually lies above the equator of the egg and in the common European frog, *Rana temporaria*, determines the plane of bilateral symmetry of the future embryo. The effect of this which is most easily visible appears directly opposite the point of sperm entry (Fig. 10): pigment granules begin to flow towards the centre of the egg and towards the vegetal pole. Yolky material drifts towards the animal pole (as it must do, if the egg is to remain spherical) and hence the pigment is here observed through a yolky cytoplasmic film (see Fig. 10). This flowing-in of pigmented cortical material occurs mostly at 180° longitude, but does in fact extend both ways around the egg at about the level of the Tropic of Capricorn. The result is a **grey crescent** which is visible with little difficulty on fertilized frog's eggs (it is seen in section in Plate Va).

Cleavage now supervenes. The first cleavage bisects the grey crescent, the second cleavage is at right angles to this, both through the poles; the third cleavage lies above the equator, almost at the Tropic of Cancer (see Fig. 5). The cleavage is radial and proceeds around the forming blastocoele until some 64–128 cells have been produced. The grey crescent has maintained its integrity throughout, but is now, of course, distributed through *cellular* tissues. The tissue including it now begins to bulge into the blastocoele as a fold, resulting in a "smile" on the surface of the egg (Fig. 11). This is the beginning of the morphogenetic (organ forming) movements in the frog which will result in the complete embryo (p. 27).

That fold of tissue which bulges into the blastocoele now reaches towards the animal pole, but no tear appears in it. A useful analogy is the deformation of a flaccid balloon by a hand pushed into it almost tangential to the surface. If this experiment is performed, it will be seen that the corners of the groove (the original "smile") must extend around their line of latitude in order that more tissue may roll inwards over the "upper lip". In fact, in the frog's egg, the "smile" extends even further than "ear to ear" and meets around the other side of the egg. During its progress, tissue is rolling over the lip and becoming changed in the process to become primitive gut, **notochord** and mesoderm (see over). That part of

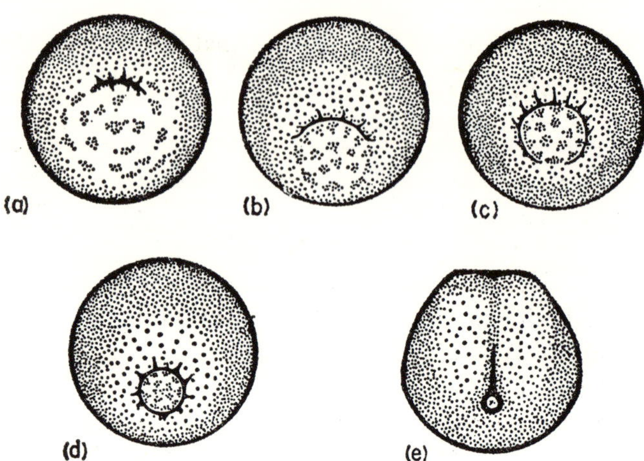

FIG. 11. *Stages in the growth of the frog blastopore lip, viewed from vegetal pole In* (*d*) *a yolk plug has formed, and in* (*e*) *the blastopore is nearly closed and the neural plate has formed.*

the lip which initially is formed (in the original position of the grey crescent) consists of tissue which will be transformed into notochord and this lip will finally represent the most *posterior* end of the *dorsal* side of the embryo; it is therefore called the **dorsal lip**. (Note that it has only the same reality as a whirlpool—its matter is constantly changing.)

The tissue rolling over the more **lateral lips,** and later over the **ventral lip,** when a full circle is complete, immediately splits into two layers. The inner layer, still, of course, in contact with the "outside world" as represented by the **enteron cavity,** will be the wall of the **gut;** the outer layer, lying between gut and that tissue which remains on the outside, will form the mesoderm. This mesoderm itself splits into two layers. One, apposed to the gut, is called **splanchnopleure*;** the other, apposed to the inner aspect of the surface tissue, is called the **somatopleure*.** The space which appears between them will become the **coelom** of the embryo and adult (see Figs. 12 and 13). The primitive gut, or enteron, extends until it almost touches the anterior wall of the gastrula, more or less where the sperm originally entered. Meanwhile, the yolky cells which were originally vegetal have passed over, or have been engulfed by, the lateral and ventral lips which

* The words are occasionaly used to mean mesoderm+endoderm and mesoderm+ectoderm respectively.

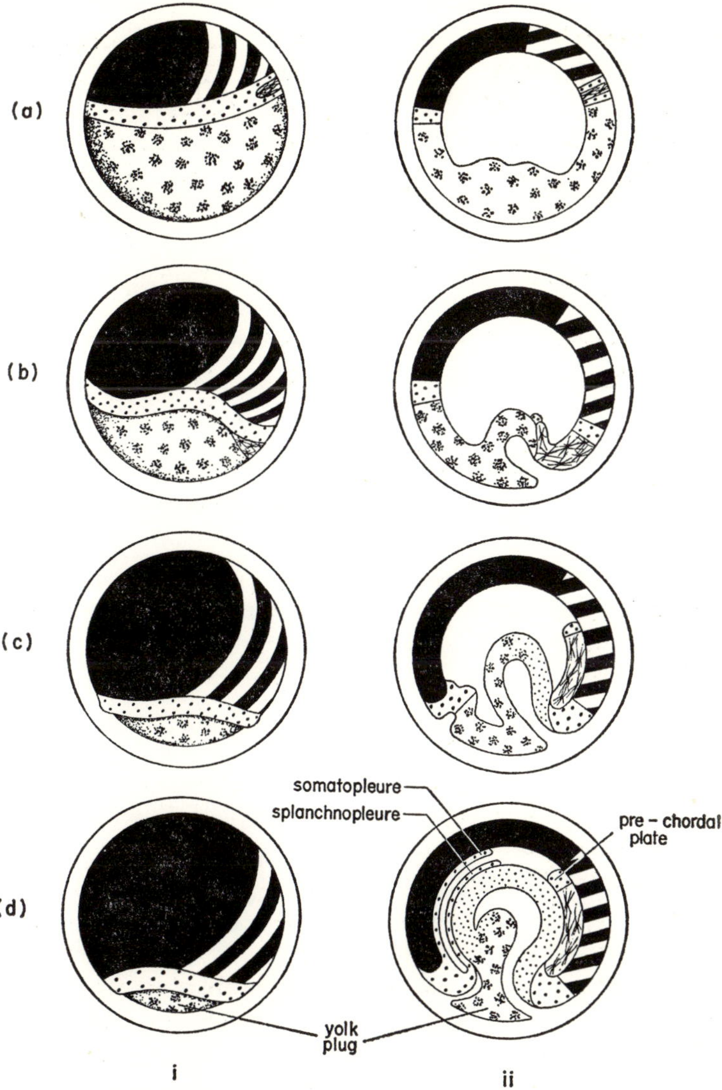

FIG. 12. *Stages in gastrulation of the frog, shown in the round* (*i*) *and in vertical section* (*ii*). (*a*) *A late blastula.* (*b*) *Early gastrula, comparable with Fig. 11* (*b*). (*c*) *Late gastrula, comparable with Fig. 11* (*c*). (*d*) *Yolk-plug stage, comparable with Fig. 11* (*d*). *In* (*c*) *the endoderm forming the gut roof develops from those cells of the vegetal hemisphere containing least yolk; these cells are sometimes difficult to distinguish from the early notochord.*

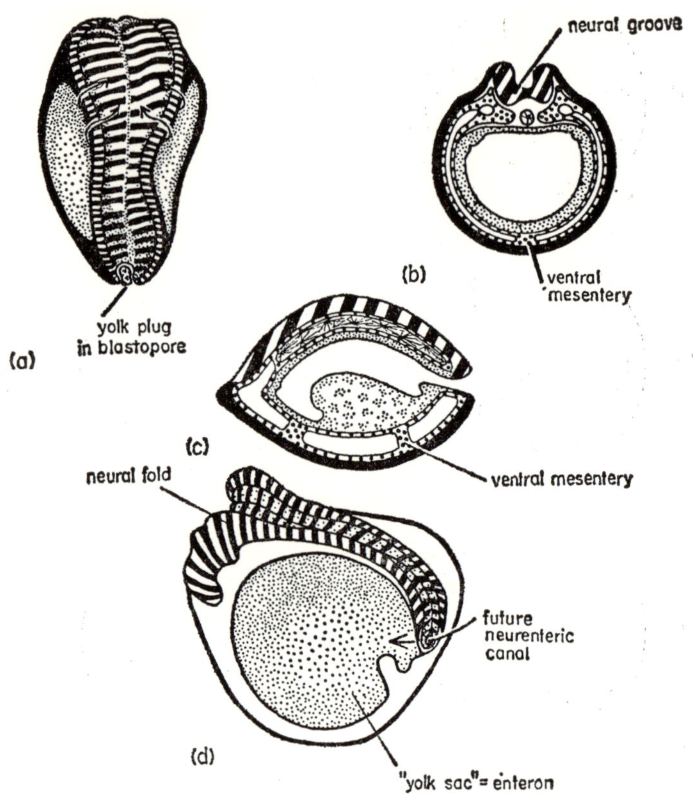

FIG. 13. *Stages in neurulation of the frog.* (*a*) *Neurula viewed from above showing the neural ridges folding over.* (*b*) *T. S. of* (*a*). (*c*) *An L. S. of a stage comparable with* (*a*). (*d*) *A slightly later stage illustrating the establishment of a neurenteric canal at the site of the "old" blastopore* (*Compare Fig. 14* (*f*).

mark the edge of the enteron. This boundary to the enteron (which must again be noted to have only the same reality as a whirlpool) is called the **blastopore.** This blastopore was seen to consist firstly only of the dorsal lip, later of lateral and then ventral lip as well. It should be realized that once the blastopore forms a complete ring, it can only *decrease* in circumference as more tissue rolls over its lip into the inside, until finally it has decreased to a small posterior (and vegetal) ring which bounds a mass of yolk which has not yet been completely engulfed, the **yolk plug** (see Figs. 11 and 12).

As that tissue which rolls over the dorsal lip transforms into notochord, it has a peculiar action upon the tissue lying immediately above it; this tissue, still on the outside, is changed so that it will become neural (nervous tissue). Because the movement of tissue near the dorsal lip is in the bilateral plane of symmetry of the embryo, the neural tissue appears as a longitudinal plate, also symmetrical in this plane.

In the area of the notochord and this **neural plate,** tissue tends to move from the more lateral towards the midline of the animal. The more dorsal region of the gut (enteron) becomes long and narrow (and the notochord becomes completely separated from it); the neural plate forms two parallel

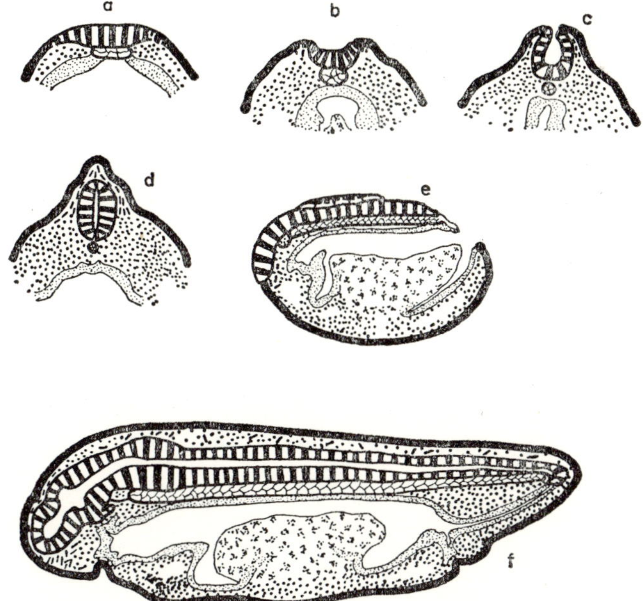

FIG. 14. *Neurulation in the frog.* (*a*) *Transverse section of upper part of young neurula.* (*b*) *The neural folds are appearing, and neural crest cells are dropping off.* (*c*) *The neural groove.* (*d*) *The neural tube has sunk below the surface.* (*e*) *Longitudinal section of a stage between* (*c*) *and* (*d*). (*f*) *Longitudinal section of a stage comparable with* (*d*). (*Compare Fig. 13.*)

ridges which then arch over the groove which appears between them (Figs. 13 and 14). It will be seen that in this way a deep groove is produced, which is soon overgrown by ectoderm and hence "drops into" the embryo as a thick-walled tube running from anterior to posterior. This separates completely from the skin to which it was attached, which then heals over in the dorsal midline. All of that tissue which was **induced** to become neural under the **inductive influence** of the notochord underlying it was concerned in the formation of this **neural tube.** At the most anterior end of the tube, its cavity remains for a while continuous with the outside world via the **neuropore,** while at the posterior end a most interesting phenomenon occurs. The appearance of neural folds and neural groove proceeds posteriorly until it reaches the blastopore, which is almost closed. The neural folds extend on either side of the blastopore recruiting material from the lateral lips, and then arch over and fuse. In this way, the lumen of the developing gut loses its connection with the outside world except through the length of the neural tube and via the neuropore. The region of junction of the enteron and the neural tube is called the **neurenteric canal** (Fig. 13).

The tail now begins to make its appearance. Immediately anterior and dorsal to the old position of the blastopore, now roofed over as the neurenteric canal, cell division and cell movement in all the tissue results in a bump on the surface. This bump projects over the rear end of the animal, and the rear end of the neural tube projects into it; on either side of this, are mesodermal extensions from either side of the notochord; the whole structure is covered with ectoderm. The posterior end of the notochord, which is still relatively undifferentiated, now extends into this process. The whole structure now grows, until it equals the length of the rest of the embryo. The mesoderm breaks up into segments which subsequently differentiate into muscles and become innervated from the neural tube next to them; thus the tail starts twitching.

The embryo that we have now produced is more or less comparable with the longitudinal section shown in Fig. 14 (and with the generalized vertebrate of Fig. 9) and in the next chapters we will consider the early development of other animals. However, before going on, a word of encouragement is in order here. The morphogenetic movements which we have just described in the frog *are* hard to visualize, because one has to try to think in three dimensions aided by diagrams which obstinately remain two-dimensional. But, once the processes involved in gastrulation are grasped, most of the rest of embryology becomes easy to follow and understand. So, stick with us!

Before considering development in other animals, mention should be made here of the **germ layer theory.** In the frog, as in the great majority of animals, the process of turning a hollow ball of cells into a baby animal involves the formation of three basic layers of tissue, an outer skin **(ectoderm),** a middle layer **(mesoderm)** in which are muscles, blood system and most of the substance of the animal, and an inner layer **(endoderm)** forming the functional epithelium of the gut. These layers, in general, maintain their integrity from the gastrula onwards and used to be thought to constitute three systems, each with its characteristic organs (the tissues were called **germ layers**). The differences between ectoderm, endoderm and mesoderm are now considered to be secondary, resulting from such interactions as occur at gastrulation and not to be primary, determining the form of the animal. However, the terms are in general use and, provided they are used only for descriptive purposes, can be helpful and not too misleading. We have used "germ layer" form in our colouring of Fig. 9, etc. We could have coloured Fig. 9 in any purely arbitrary way (e.g. in transverse slices), but the system using "germ layers" is more successful here for just the same reason that the germ layer theory itself was a popular view of development for over 50 years. Large discrete areas of the blastular surface move to form each of these germ layers, which themselves maintain their continuity to a large extent. Therefore, a fate map constructed in this way is *simpler* than that derived by colouring, for example, transverse slices of the formed embryo. It nevertheless gives as much information.

We will now consider the early development of other animals prior to returning to the story of further vertebrate development; by this approach, we hope to put vertebrate development more obviously into the context of animal development as a whole.

NEMATODES

As has already been mentioned (p. 19) nematodes, rotifers and their relatives differ from other animals in that their development is extremely determinate, even to the number of cells in each organ. Furthermore, this determination appears, at least in the nematodes, at a very early stage. This, together with the amoeboid nature of the sperms, makes it probable that they are not closely related to other Metazoa. We will, therefore, consider their development separately and avoid comparisons. Figure 15 shows some early stages in the development of nematodes (see p. 150 for techniques for obtaining and viewing living stages of nematode develop ment, in *Rhabditis*).

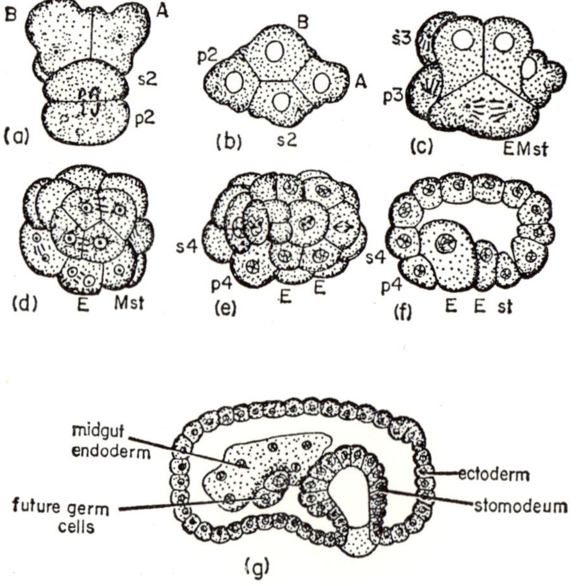

FIG. 15. *Stages in the development of a nematode. Development of the nematode* Parascaris equorum *(from Hyman, after Boveri 1899). (a) T shaped 4-cell stage (b) Later 4-cell stage, rhomboid in shape. (c) 7-cell stage. (d) 18-cell stage. (e) Ventral view. (f) Sagittal section through the "gastrula". (g) Sagittal section showing the formation of the stomodeum.*

The eggs pass down the oviduct and meet the amoeboid sperms, which stick to the egg membrane. The sperm passes into the egg and a very thick shell is secreted. Many nematodes, including *Rhabditis*, are viviparous and all stages can be seen inside. The 4-blastomere stage is initially T-shaped. The cells then rearrange to form a rhomboid (Plate IV e, f); this takes about 15 min in *Rhabditis*. All divisions are absolutely fixed in direction and timing. This must be the case when the organization of the egg has already been achieved before cleavage starts; the various parts all take prescribed paths to their final positions.

Those cells which produce the internal organs (midgut, gonads and "mesenchyme") are all forced into the interior by directive cell divisions. Although Plate IVg and Fig. 15 (g) have a certain superficial resemblance to gastrulae of other animals (for example echinoderms), there can be no comparison in embryological terms. Neither flow of tissues into new relationships nor interaction resulting in greater biochemical complexity appears to result from the cell rearrangements of nematodes. The already different cytoplasms are simply supplied with nuclei and pushed into special positions. (In the rotifer embryo the cells are also highly determined, but they do move considerably to arrive at their final organization; here again, no interactions seem to occur between them.) When all the parts have come to their final positions (Fig. 15 and Plate IVg), the embryo elongates into a miniature adult and then hatches. No more cell divisions are thought to occur after hatching, except in the ovary and testis and perhaps in the epidermis. This sheds a series of cuticles during the growth of the larva.

The development of the nematode *Caenorhabditis elegans* is probably the best known of any animal. It seems a pity that so much effort has been put into investigation of such an aberrant embryology, but of course only this kind of development *can* give such consistent and regular results.

POLYCHAETES

The eggs of polychaete worms are homolecithal to mesolecithal. The development may best be followed in homolecithal types, as the presence of large quantities of yolk modifies the development to a greater or lesser extent in other forms. Plate IV shows the egg of a *Nereis* before fertilization and at three stages in the elevation of the fertilization membrane. It will be observed that considerable organization of the egg cytoplasm occurs and that the yolk aggregates near, but not at, the vegetal pole. After this organization has occurred cleavage then proceeds as follows (see Fig. 16). This is typical for both annelids and molluscs. Its aim is to achieve cellularity without destroying this organization.

The first cleavage is vertical and divides the egg into right and left halves. In forms like *Pomatoceros* (Plate II), where a red pigment has been localized into a small area opposite the point of sperm entry, the cleavage plane may be seen to divide the pigment patch into large and small portions. The second division is also vertical, but at right angles. We now have four cells lying approximately in the same plane (actually two touch at the animal pole and the other pair touch at the vegetal pole). At this stage we may pause to consider the distribution of certain "organ-forming substances" (see p. 20) in the egg. The aggregated pigment and much of the yolky material comes to lie mostly in one of these cells. In many forms, especially among the molluscs, this area of cytoplasm is extruded as a **polar lobe** just before each division and then withdrawn into the appropriate cell when the division is complete. Because the cells are already different, it is convenient to label them and their descendants (Figs. 16, 17 and Table 1); this is not the case, of course, with indeterminate forms.

The four cells, viewed from the animal pole and reading in a clockwise direction, are called A, B, C, D. D is often the largest and is that cell which contains the most pigment and yolk; these mark the cytoplasm which will form the adult mesoderm and ectoderm (muscles, gonads, etc. and skin). The next division (which is spiral, see p. 25) results in four smaller cells lying above (animal to) and alternating with four larger cells

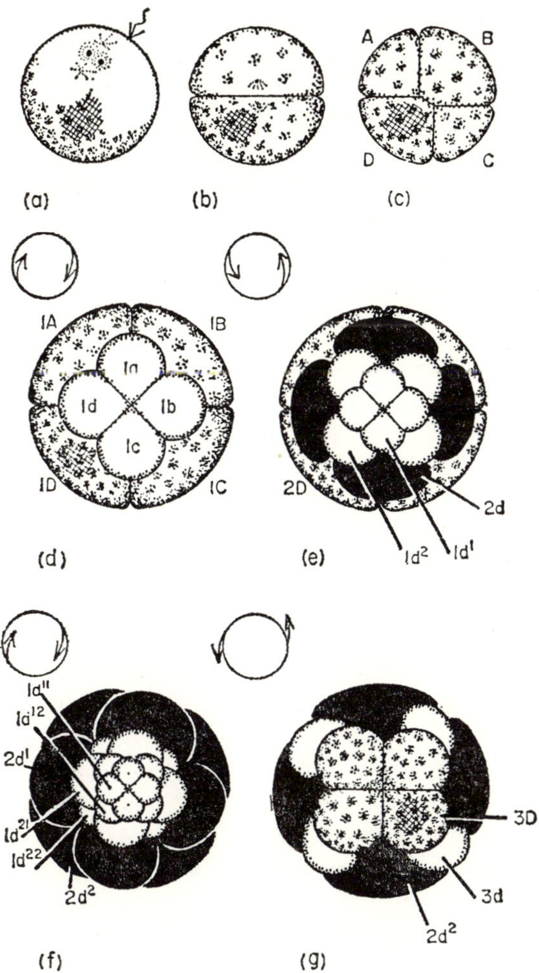

FIG. 16. *Early spiral cleavage stages. (a) Fertilized egg with arrow indicating point of sperm entry. (b) 2-cell stage. (c) 4-cell stage. (d) 8-cell stage; the 4 micromeres alternating with the 4 large yolky macromeres. Viewed from the animal pole. (e) 16-cell stage from the animal pole. (f) 3-cell stage viewed from the animal pole. (g) 32-cell stage viewed from the vegetal pole.*

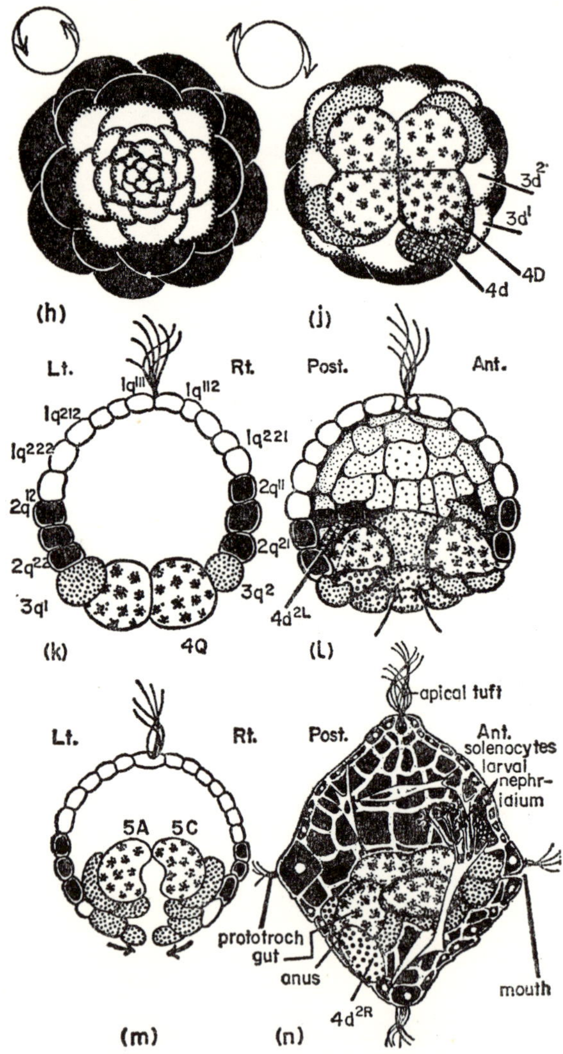

FIG. 17. *Spiral cleavage stages (continued). (h) 64-cell stage from the animal aspect. (j) 64-cell stage from the vegetal aspect. (k) Vertical section of a 64-cell stage (from left to right). (l) Vertical half (anterior to posterior) of a later stage showing invagination of the micromeres to form gut, i.e. section from B—D quadrant. (m) Vertical section (taken at right angles to the plane of section in (l),*

TABLE I. Spiral cleavage nomenclature

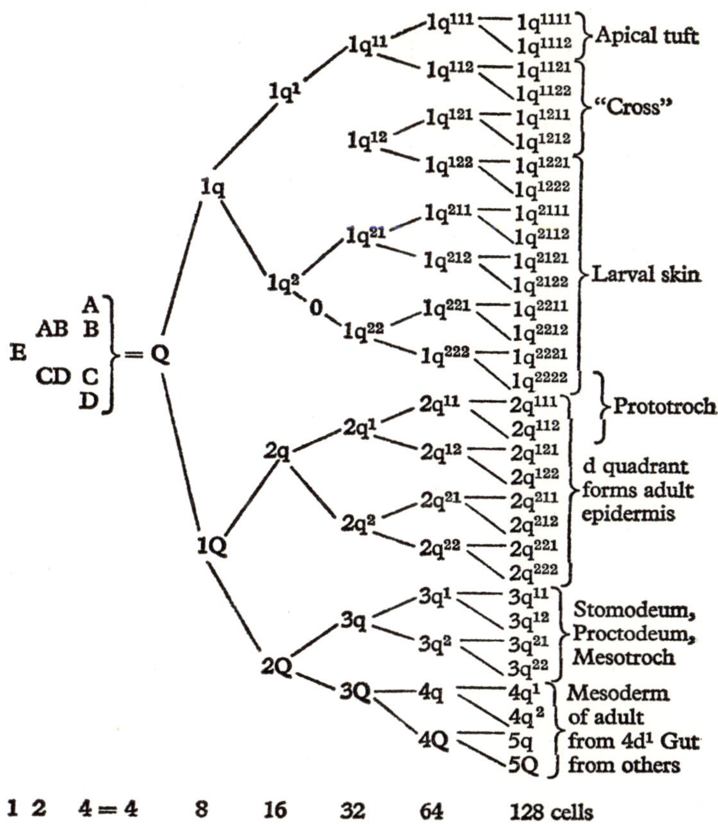

---

and reduced) showing how the floor of the gut is formed, i.e. from A quadrant to C quadrant. (n) Transparency of a trochophore with mouth at the site of the front of the "blastopore" and anus perforated at the D end of the "blastopore". Note the 4d mesodermal cells lying each side of the gut.

(Plate VIb). The upper quartet are called **micromeres** (1a–d) and the large vegetal quartet are called **macromeres** (1A–D). Little of the pigmented material or yolk has been lost from D into 1d, but most remains in 1D. The division which produced the first quartet of micromeres was, viewed from the animal pole, angled in a clockwise direction, and hence 1a lies clockwise to 1A, 1b lies clockwise to 1B, etc. The next division, to attain sixteen cells, is anticlockwise and all eight cells divide together; 1a, 1b, 1c, 1d, or in general 1q (a **quartet** of cells) divide to give $1q^1$ and $1q^2$, $1q^1$ being more animal. 1Q divides to give 2Q and 2q, 2Q being more vegetal; 2q are the second quartet of micromeres. The next division, to give thirty-two cells, is again clockwise, producing the following quartets: $1q^{11}$, $1q^{12}$, $1q^{21}$, $1q^{22}$, $2q^1$, $2q^2$, 3q and 3Q. This nomenclature is shown in Table 1. The next division, to produce sixty-four cells, is usually the last to be synchronous; it is anticlockwise. The quartets are now $1q^{111}$, $1q^{112}$, $1q^{121}$, $1q^{122}$, $1q^{211}$, $1q^{212}$, $1q^{221}$, $1q^{222}$, $2q^{11}$, $2q^{12}$, $2q^{21}$, $2q^{22}$, $3q^1$, $3q^2$, 4q, 4Q. At this division, most of the pigmented material may be seen to be localized in cell 4d, although some of the other "active" material from macromere D has already been lost into micromere 2d and its descendants. The four macromeres 4Q are extremely yolky and destined to form part of the gut of the embryo and adult; very often they are delayed in their next division relative to the other cells. It will be noted that the cleavage is not truly spiral but rather "alternating". This results in the keying together of the cells of each **quadrant** (a quadrant is the product of one blastomere of the four-cell stage). This ensures that each part of the original egg cytoplasm maintains its position during cleavage; it is positioned around the blastocoele as it originally lay around the egg nucleus. Even more important than the spatial organization of cytoplasm is that of the egg cortex. Experiments have shown that abnormality, which is almost always fatal during cleavage, results from any change in the cortical architecture.

The next division, to form 128 cells, is again clockwise in most of the cells, but by now the ordered system is beginning to break up. The sequence of cells animal to vegetal is now:

$1q^{1111}$, which bears an **apical tuft** of cilia;

$1q^{1112}$, which may also have cilia;

$1q^{1121}$, $1q^{\cdot 1122}$, $1q^{1211}$, $1q^{1211}$, which form part of a pattern of a cross which differs slightly in annelids and molluscs and may enable the embryos to be distinguished.

$1q^{1221}$, $1q^{1222}$, $1q^{2111}$, $1q^{2112}$, $1q^{2121}$, $1q^{2122}$, $1q^{2211}$, $1q^{2212}$, which form much of the **larval skin.**

$1q^{2221}$ and $1q^{2222}$ which may bud off cells from their inner aspect into the blastocoel to produce **larval ectomesenchyme** (i.e. the "stuffing" of the larva) as may other cells descended from the first and second quartets (Fig. 15).

$1q^{2211}$, $1q^{2212}$, $1q^{2221}$, $1q^{2222}$ bear long cilia. Some cells from the second quartet, $2q^{111}$ and $2q^{112}$, may also bear cilia. These ciliated cells form a girdle around the equator of the embryo, now called a **trochophore** or **trochosphere;** this girdle is called the **prototroch** (Fig. 17). Descendants of 2d will extend to form almost all of the adult epidermis (Fig. 18).

The third quartet, $3q^{11}$, $3q^{18}$, $3q^{21}$, $3q^{22}$, may contribute in part to the junction of the gut with the skin at mouth and anus, the **stomodeum** anteriorly and **proctodeum** posteriorly, and some may later develop cilia to form a more vegetal girdle, the **mesotroch.**

The fourth quartet, 4q and 4Q (or sometimes $4q^1$ and ^{2}, 5q and 5Q), begin to roll into the blastocoele and so create a groove on the vegetal aspect of the embryo. The sides of this groove are $4a^1$ and ^{2} (left), $4c^1$ and ^{2} (right), while the anterior is formed by $4b^1$ and ^{2}, and the posterior end by $4d^1$ and ^{2}. $4d^1$ divides into two cells lying *left* and *right* of the posterior end of the groove, and projecting up into the blastocoele, $4d^{1L}$ and $4d^{1R}$ (Fig. 17, (l), (m) and (n)). The groove now closes as 5A and C meet in the midline, forming the gut, and its posterior end may become the anus. (Understanding of these events will be greatly aided by the making of plasticine or wax models.)

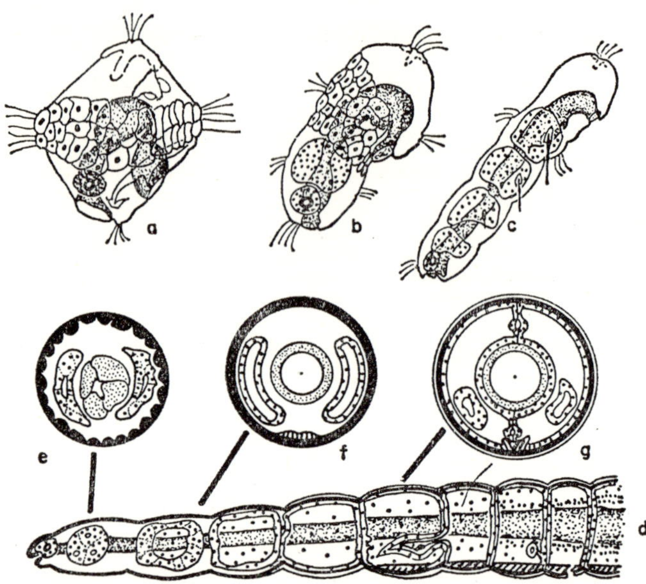

FIG. 18. *Transformation of the trochophore larva into the worm.* (*a*) *The fully formed trochophore.* (*b*) *4d has budded off one segment, and the descendants of 2d have extended to form a saddle shape; their outlines are shown.* (*c*) *Four segments have been formed and the gut is extended.* (*d*) *Very diagrammatic view of the rear end of an older worm.* (*e*), (*f*), (*g*) *Transverse sections of the segments indicated. In* (*e*) *the coelom is appearing, in* (*f*) *the mesodermal bubbles are growing and the nervous system is appearing, and in* (*g*) *the bubbles have met dorsally and ventrally, forming the blood vessels; ovary or testis sac is also seen.*

The above description of the events leading to the trochophore larva apply with little modification to all of the polychaete worms and to many of the molluscs which have homolecithal eggs. The transformation of the trochophore into the juvenile worm is a very strange process. Most of the structures of the trochophore do not contribute to the adult—they play their part in the life of the larva as it swims actively in the plankton. Locomotory organs are the apical tuft, usually leading, and the prototroch; the stomodeum becomes filled with phytoplankton (diatoms and other algae) as they are caught in the turbulence of the prototrochal cilia, and it injects this mass into the stomach where it is digested. The excretory organs may be working in the elimination of nitrogenous waste, but it is probable that in so small an organism this is mostly taken care of by diffusion.

Let us consider the transformation of the trochophore into the familiar worm (Fig. 18); almost all of the substance of the latter is produced from the descendants of *two cells only*, 2d and 4d. Cells $4d^{1L}$ and $4d^{1R}$ are now sitting one on each side of the posterior end of the gut. These two cells produce clusters of cells which move up on either side parallel to the gut and develop cavities, the coelom, amongst themselves. During their growth the cavities swell until left and right clusters come into contact above and below the gut (Fig. 18). This series of **mesoderm buds** on each side resembles a chain of soap bubbles. The blastocoele is almost obliterated but may survive in part as the dorsal and ventral blood vessels, trapped between the mesodermal "bubbles" of each side. The front and rear walls of the "bubbles" will form the **septa** and the "bubbles" form the **segments.** The production of these mesodermal segments initiates an increase in length (animal-to-vegetal in the egg *and* mouth-to-anus in the embryo); the skin compensates for this by division of the descendants of 2d, which increase their area to a saddle shape and then meet ventrally. The original embryo now only persists as a **prostomium** surrounding the mouth and lying above it and a **pygidium** surrounding the anus and lying above it. The whole of the middle length of the worm except foı these tiny ends is provided from division of 2d (the **first somatoblast**) and 4d (the **second somatoblast**); the gut has elongated by division of the fourth and fifth quartets of micromeres rather than by the division of the macromeres; 4d is also called the **primary mesoblast.**

During the embryology of the earthworms, however, there is no ciliated larva. This reduction would be expected in a land animal. Correlated with this is a simplification of the embryology, although the cleavage pattern is obviously derived from spiral cleavage. This is shown diagrammatically in Fig. 19; the embryology of *Lumbricus trapezoides*, illustrated here, is of further interest, because each egg results in *two* embryos side by side.

It must be emphasized that the story of worm development is mostly a story of distribution of already organized material. The most dramatic epigenetic activity has occurred once cleavage has started.

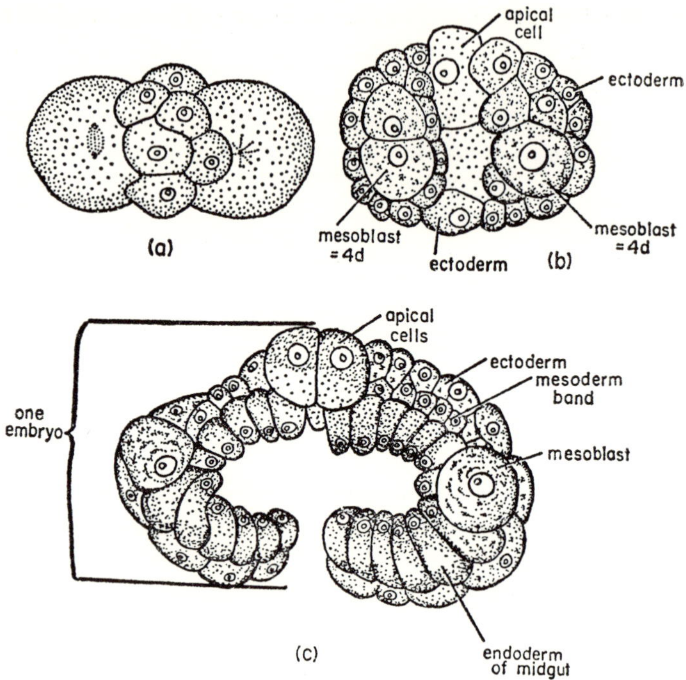

FIG. 19. *Development of the earthworm* Lumbricus trapezoides (*from Kleinenberg 1879*). *Each egg gives 2 embryos.* (*a*) *An early cleavage stage: the spiral arrangement of the micromeres is obvious.* (*b*) *The embryos have begun to develop independently and the spiral form has been lost.* (*c*) *A later stage in vertical section, corresponding to 2 trochophores with their ventral sides apposed.*

MOLLUSCS

The early development of most molluscs resembles that of polychaete annelids very closely. The trochophore, present even in many freshwater snails as a stage passed in the egg mass, transforms into a **veliger** larva. This often has enormous ciliated **velar lobes** (Fig. 20 and Plate VIf) which assist in food collection and swimming. The veliger larva of gastropods undergoes **torsion,** the visceral mass rotating over the head and foot, bringing the mantle cavity and anus over the head (see Plate VIf and Fig. 20). Many freshwater gastropods, for example *Limnaea* and *Planorbis*, show stages closely comparable with Fig. 16. They all have determinate

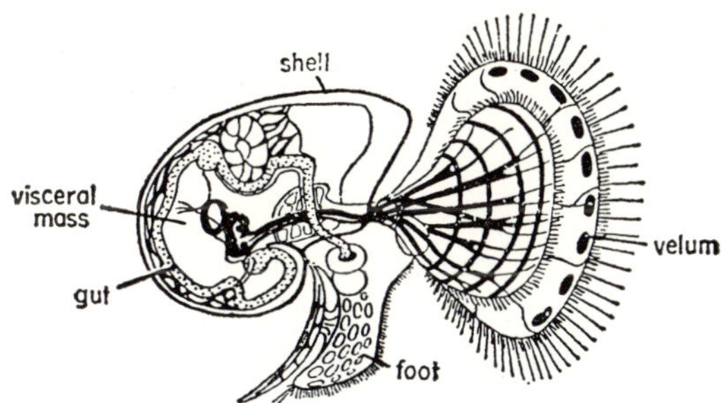

FIG. 20. *Diagram of a veliger larva of a gastropod mollusc. This results from a trochophore similar to Fig. 17 (n) and the early form of the young snail can clearly be seen.*

eggs and spiral cleavage. It has been shown, however, that in *Limnaea* some epigenetic events occur as late as the veliger. The bulge of gut towards the animal pole of the egg causes development of the rudiment of the **shell gland** from the descendants of the first and second quartets of micromeres. If the gut invagination is prevented or retarded, the shell gland either does not develop or may develop in the abnormal area of contact. This is our first clear example of **induction** of an organ in an invertebrate. The gut induces the ectoderm to form the shell gland.

Other molluscs, the bivalves and cephalopods, develop differently. The bivalves usually have spiral cleavage, trochophores, then a two-shelled "veliger". *Anodonta*, the freshwater Swan Mussel, keeps the early eggs and larvae in its gills. The late larva has a strange pair of hooked shells with which it "hitches a lift" on the skin of passing fishes, later dropping onto the mud to become free-living.

The egg of the squid, *Loligo peali*, is very determinate with a complex architecture; cleavage of the blastodisc (the eggs are telolecithal, like those of all cephalopods) results in a plate of cells which organize to form the head and tentacles—encroachment of the edge over the yolk results in the visceral mass (see Plate VII d). It is very difficult to make comparisons with veligers, just as it is difficult to compare insect development with that of the crustaceans (p. 52) or indeed that of teleosts with other vertebrates (p. 61).

ARTHROPODS

The early development of the crayfish is shown diagrammatically in Fig. 21. After this peculiar cleavage, when the surface of the egg is covered by cells and the centre filled by yolk (a stage perhaps comparable to the blastula), a groove appears ventrally which invaginates to form gut. Despite the supposed relationship of arthropods and annelids, it is very

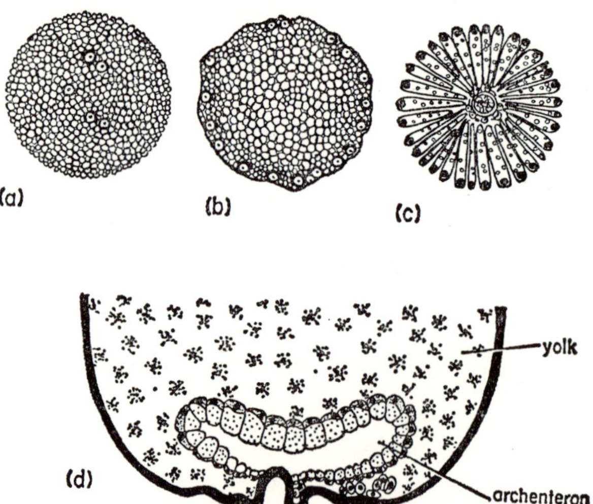

FIG. 21. *Stages in the development of the Crayfish* Astacus fluviatilis. (*a*) *The nuclei have commenced to cleave.* (*b*) *They have taken up a position just under the surface of the yolk filled mass.* (*c*) *Each has acquired a segment of the yolky cytoplasm.* (*d*) *A section of a much later stage. The embryo is formed by only part of the surface*

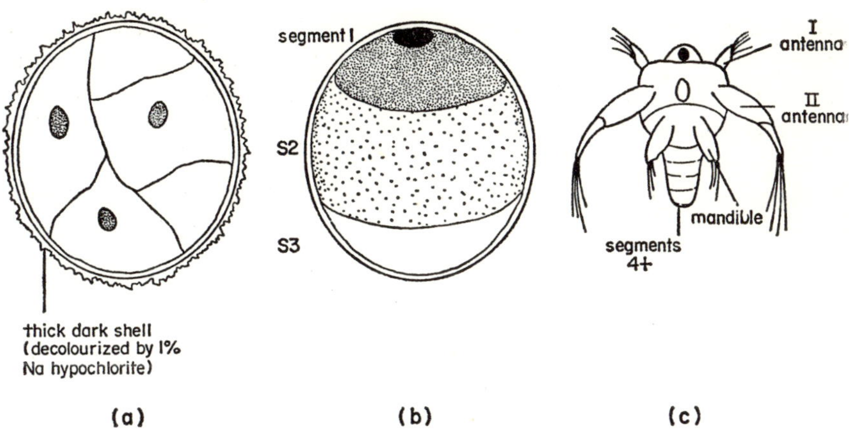

FIG. 22. *Development of* Artemia.

difficult to establish any comparisons between this kind of development and spiral cleavage. The mesodermal bands do not appear to be produced from two separate cells but seem to arise from tissue interactions during the formation of the gut. It is probable that in the arthropods *some* of the epigenetic processes have been delayed until after cleavage (like the induction of the shell gland of gastropods) and experimental division of the egg at early stages confirms this.

Although crayfish will sometimes breed in captivity there are many other Crustacea whose development can be more readily studied. Of these the brine shrimp, *Artemia*, is certainly the most convenient; but the little copepod, *Cyclops*, is a useful alternative, as the eggs are carried around by the female while they develop. *Artemia* and *Cyclops* eggs cleave in much the same peculiar fashion as we have described for the crayfish and the cellular, yolky mass becomes organized into three parts, of which the first occupies half of the egg, the second forms a disc under this, and the third is the other pole of the egg. These will be the first three **segments** of the animal and soon develop stumpy appendages. The largest segment develops a pair of antennules and a dark, median eye as nerve cells differentiate. The second (disc-shaped) segment develops the antennae, and the third segment develops projections which will become the mandibles (Fig. 22). Both *Cyclops* and *Artemia* hatch at this **nauplius** stage (Plate III l, m) and moult, without feeding, to produce a **metanauplius** with one more segment and pair of limbs; it also can be seen to have a knob on the back end from which further segments will be produced at successive moults.

The development of insects is altogether aberrant, but can be seen to be derived from systems comparable with that of crustaceans. The lozenge-shaped egg contains two specialized areas of cytoplasm, the **fertilization centre** containing the nucleus towards one pole of the egg and the pole plasm at the other pole. Migration of the nucleus during and after

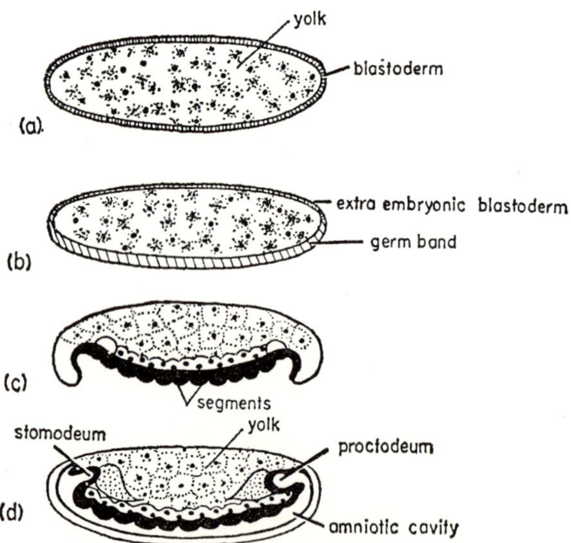

FIG. 23. *Stages in the development of an insect.* (*a*) *Cleavage of the nuclei in the centrolecithal egg. One layer of nuclei has already recruited cytoplasm to make a blastoderm at the surface.* (*b*) *Under the influence of the activation centre the blastoderm thickens in the region of the future embryo.* (*c*) *The germ band is segmenting and sinking. The yolk has now cleaved and a true mesoderm has appeared.* (*d*) *After fusion of the amniotic fold; the embryo lies in an amniotic cavity.*

fertilization, and interaction between the nucleus (or daughter nuclei) and the activated cytoplasm have been shown to be necessary for organization of the embryo (Fig. 23). It appears that the development of insects ranges from indeterminate to a degree of determination approaching that of the nematodes, e.g. the gall midges (*Cecidomyidae*). These insects are very useful experimental material. By inactivation of the nucleus or parts of the cytoplasm with X-rays or ultraviolet rays, it has been shown that until the nuclei migrate into the specialized area of cytoplasm (lying in a thin film over the central yolk), they are not restricted in their ability. They can contribute to any part of the embryo; they are totipotent (p. 70). However, once they have divided in the already organized cytoplasm, they lose their totipotency and can only continue further along the same road; they can never turn back and take another turning (see p. 140). Only those nuclei which have migrated into the pole plasm will contribute to eggs and sperms. A mutation in *Drosophila* (grandchildlessness) causes females homozygous for it to produce eggs lacking pole plasm; they make normal but sterile progeny.

Insect embryology is a very involved subject; for more details, the reader is referred to more specialized works. Hemimetabolous insects may be regarded as having a succession of larval stages (the **nymphs**) which have great advantages in growth studies, because the successive casts provide a permanent record of the history of the organism. Holometabolous insects show a very complex metamorphosis (p. 145) from the larva into the adult; this is one of the best documented transformations in the animal kingdom, but has usually been treated from the endocrinological point of view rather than the embryological. Nevertheless, this subject should be in the repertoire of any competent developmental biologist.

ECHINODERMS

The eggs of echinoderms are perhaps the favourite experimental material of those workers concerned with fertilization and early development, as they are relatively easy to obtain and to culture. Fertilization has been studied in detail and most of our information about fertilization is derived from such work. The presence of a planktonic larva in almost all forms means that the early embryology leads to this larva (Plate VIg) rather than to the adult. The metamorphosis of the larva into the adult varies from species to species and cannot be considered here.

During fertilization a space appears between cortex and vitelline membrane (comparable to that shown in Plate IVb and d). This process is associated with the breakdown of cortical granules; the membrane lifts to become a fertilization membrane. It is into this space that the second polar body is extruded and the position of its extrusion marks the animal pole. The eggs are usually homolecithal and the events following fertilization can only rarely be correlated with pigment or yolk movement. All are relatively indeterminate.

The first cleavage is again vertical and passes through the point of sperm entry in most eggs. The second cleavage is also vertical, at right angles. The third cleavage is horizontal and lies at about the plane of the equator. Note that this results in two quartets of cells lying *directly* one above the other. This is radial cleavage (Fig. 8(a) and Plate VIa). Each of these cells, if separated, can probably give rise to a whole embryo, although the extent of this indeterminacy varies. Usually, cleavage is synchronous up to about the 64-cell stage and then becomes irregular, forming a blastula around a large blastocoele.

Now gastrulation commences (see also p. 26, where a simplified account is given); at the point opposite to the original position of sperm entry the wall of the spherical blastula becomes thickened (Fig. 24(a), Plate VIc). It is extremely probable that the cytoplasm of the cells involved differs from the rest of the cytoplasm as the result of the establishment of a cortical difference before cleavage. The inner aspect of this thickening buds off small cells into the blastocoele, which will form the **larval ectomesen-**

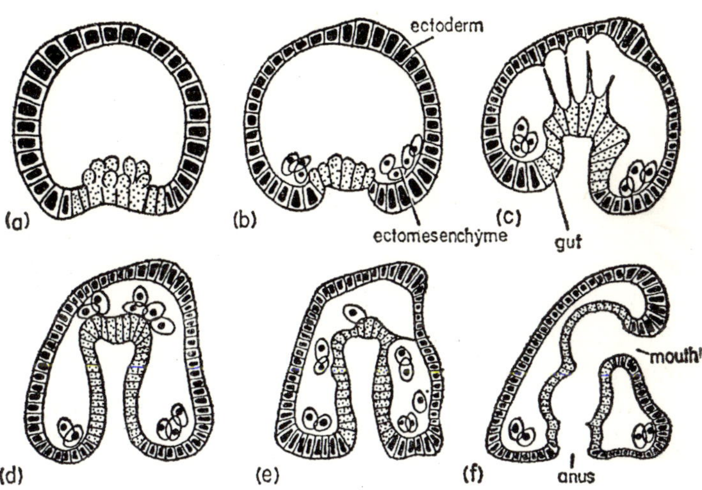

FIG. 24. *Gastrulation in echinoderms (after Gustavson and Wolpert 1961). (a) Cells of the blastula opposite the point of sperm entry change their surface characters so that they bulge into the blastocoele. (b) Ectomesenchyme cells appear which crawl about inside the blastocoele. (c) These cells take up definite positions on the wall and the gut invaginates itself hauling in on pseudopodial processes. (d) The process is nearly complete. (e) (f) Formation of the mouth.*

chyme. These cells throw pseudopodia across the blastocoele which anchor and then contract, pulling the cells into the blastocoele. Here they form a pattern on the inside wall while the area they left flattens again. Then other cells push out processes which attach to the far wall. When they contract, the ventral aspect dimples into the blastocoele (Fig. 24(b), Plate VId). Finally about a third of the blastula surface has been pulled into the rest; this forms the **archenteron,** which will produce the gut and other internal organs. The embryo is now a gastrula (Fig. 24(d)). The rim of the cup-shaped gastrula, the blastopore, which contracts like the mouth of a string purse as tissue rolls over it into the embryo, is the site of many important epigenetic events. Those cells which have rolled in have been changed, in that they can now form only gut or mesodermal organs; their action on the cells which overlie them restricts the potentiality of these overlying cells to the formation of skin and its derivative organs.

Several vesicles are now budded off from the archenteron into the blastocoele. The cavities of these vesicles will become the coelom (and the water vascular system) of the adult and their walls will produce most of the sparse adult mesoderm. Finally, various "arms", ciliary grooves and ridges and other flotation aids are produced by these embryos to assist in their planktonic distribution. Their pattern is probably dependent on that of the original primary ectomesenchyme cells, itself determined by a pattern of "stickiness" on the inner aspect of the late blastular cells, probably retained from oocyte cytoplasm structure.

The distinction may now be made between that kind of larva which results from spiral cleavage, called a trochophore (p. 45), and the very different larva of the echinoderms called a **dipleurula** (Fig. 25). The mouth

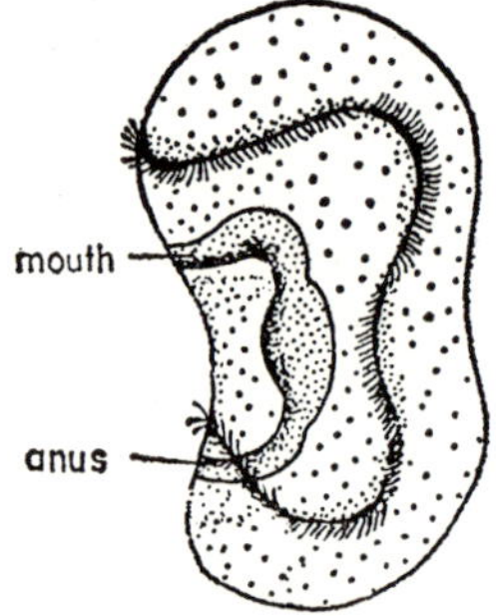

FIG. 25. *The dipleurula larva, which results from gastrulation in the echinoderms (compare Plate VIg).*

and anus of the dipleurula are both secondary; neither is the original blastopore. The mouth breaks through in the region where one side of the blastula has touched the other (i.e. near the point of sperm entry), while the anus either is derived by complex foldings from the closing blastopore or makes its appearance close to the site of closure of the blastopore. In the trochophore the mouth (stomodeum) remains as the anterior end of the ventral groove ("blastopore") during invagination of the gut, and the hind end of the groove can remain as the cloaca.

TUNICATES

This is another group which, like the echinoderms, has proved very suitable for experimental work. The eggs are determinate, but show radial cleavage. The movements of cytoplasm which follow fertilization may frequently be marked by pigments and yolk in the cytoplasm. The genus *Styela*, (also called *Cynthia*) shows this very dramatically (Fig. 7). Although no detailed description of the development of tunicates can be given, there are several points of interest which lead to informative comparisons. The group is obviously chordate, but has so many specialized features that the resemblance of the various stages is far from clear. Despite this, a lateral view of the 8-cell stage of *Styela* (Fig. 26(a)) resembles the frog blastula extraordinarily closely in the prospective fates of its areas. However, whereas the frog fate map shows only *supposed future* differences between the parts of the egg, the 8-cell stage of

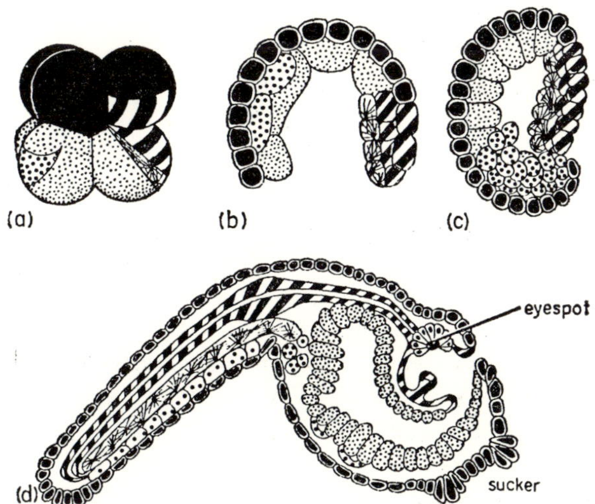

FIG. 26. *The later development of* Styela *(from various sources). (a) 8-cell stage viewed from the right side. Note that the areas are* not *a fate map, but a statement of already determined fates (except the neural area). (b), (c), (d), stages in "gastrulation" and transformation into the tadpole larva.*

Styela already has its various components restricted in their fates. This is a very good example of the close similarity between the pre-cleavage (determinate) and post-cleavage (indeterminate) morphogenetic movements in one phylum. It must, however, be noted that the nervous system of *Styela* is the result of a late induction, like the shell gland of gastropods. Later development of this form is also shown diagrammatically in Fig. 26.

VERTEBRATE EMBRYOLOGY—PREAMBLE

A description of the early development of the frog has already been given (pp. 29–37 above). The early development of the teleost fish and of the chick will now be described. All three animals will be left at a stage formally comparable with Fig. 9, the "generalized vertebrate condition". Only after dealing with the embryonic membranes of vertebrates can we return to the early development of the mammal and bring this to a comparable stage. The reason for this will become apparent.

DEVELOPMENT OF FISHES

The eggs of teleost fishes are more or less telolecithal, i.e. there is a large mass of yolk with a relatively small yolk-free **blastodisc** (Fig. 27) sitting on top of it. The relationship of cytoplasm and yolk is, however, very different from that of truly telolecithal eggs of more primitive fishes, the sharks, birds or cephalopods. At ovulation, the cytoplasm is either "mixed" with the yolk or present as a thin layer all over its surface. At laying or fertilization, the cytoplasm concentrates into a separate non-yolky mass, which alone cleaves; the yolk is entirely passive in development. In most teleosts, it appears that the plane of bilateral symmetry of

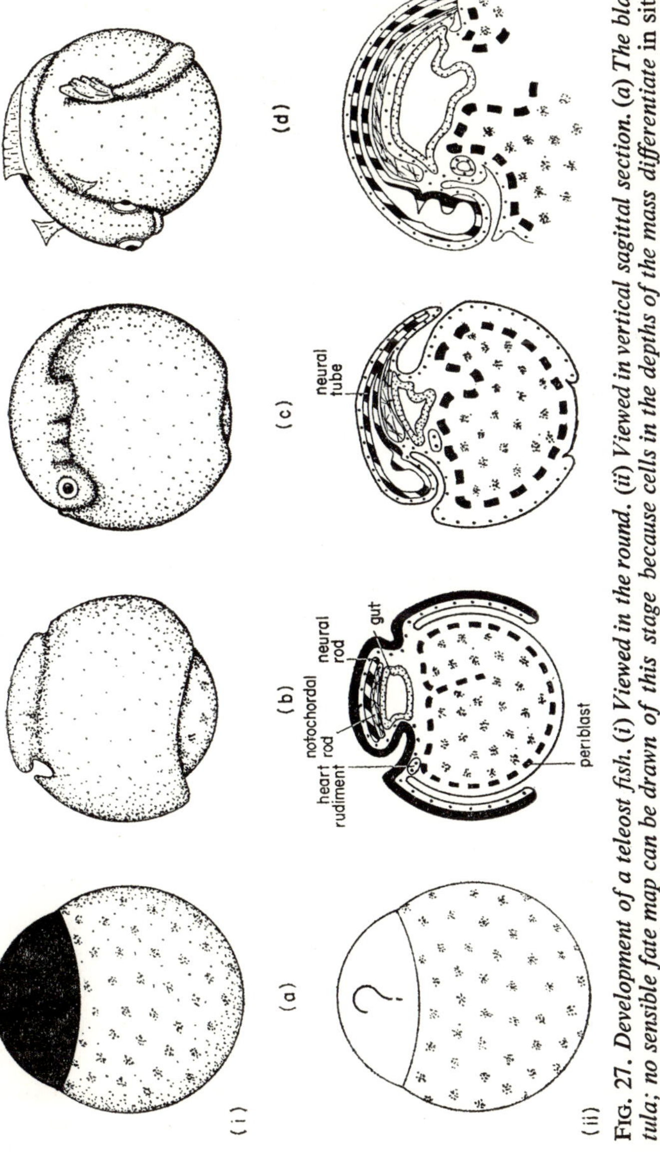

FIG. 27. *Development of a teleost fish.* (*i*) *Viewed in the round.* (*ii*) *Viewed in vertical sagittal section.* (*a*) *The blastula; no sensible fate map can be drawn of this stage because cells in the depths of the mass differentiate* in situ. (*b*) *Overgrowth of the yolk is now occurring.* (*c*) *The gut has formed a tube above the yolk and the yolk is enclosed by the periblast and lies in the* coelom. (*d*) *The young fish. Note: periblast is shown as a broken black line.*

the embryo has been determined during the production of the egg, that is to say in the ovary. The vast majority of teleosts have external fertilization, but a few forms show ovoviviparity. The guppy (*Lebistes reticulatus*) is one such form and fertilization is internal; indeed, the eggs are actually fertilized in the ovary. The anal fin of the male is rolled to form the **gonopodium,** which ejects sperms into the genital aperture of the female. The sperms may be stored for a considerable time in the genital tract (the hollow ovary) of the female and one mating may provide sperm for about ten batches of 10–30 young. The sperms penetrate the vitelline membranes of the eggs which form the inner aspect of the hollow ovary and one of them unites with the egg nucleus which lies in the blastodisc. Many other sperm nuclei, with **asters** (tubulin arrays like half-mitotic-spindles) occur around the blastodisc and in the yolk, i.e. polyspermy is the rule; this is usually the case with large eggs. Union of one sperm pronucleus with the egg nucleus causes the others to degenerate. Guppy eggs, although easier to obtain than zebra-fish eggs, are not as easy to observe microscopically, and this section will describe zebra-fish eggs (see p. 154 for techniques of obtaining and observing them).

The first divisions of the blastodisc are quite vertical and only after at least three such divisions is there a chordal (or, by comparison with the frog, horizontal) division. We would expect that this division would show us the site of the future blastocoele, but opinions have in fact been divided as to where the blastocoele lies in the fishes. Successive divisions, both vertical, horizontal and irregular, result in a situation comparable to the blastula of the frog, but whose blastocoele is only nominal (see Fig. 27 (a) and Plate IX) and with all the yolk *outside* the cleaving cells. The "gastrulation" movements are at first sight rather different from those of the frog. The circumference of the blastoderm extends outwards over the surface of the yolk (see Fig. 27 (b)) and tissue may roll over its edge, all round. If tissue does indeed roll over, then these edges of the blastoderm are obviously comparable with the lips of the frog blastopore. Engulfment of the yolk is then comparable to the engulfment of the vegetal yolk-containing cells of the frog by the lateral and ventral lips of its blastopore. Naturally, these lips of the fishes take much longer to engulf the yolk than do those of the frog; indeed, the embryo is sometimes well formed before the yolk is completely covered. As the blastoderm extends around the yolk, a thickened ridge of cells builds up in the future embryonic axis (Fig. 27 and Plate XI). Probably without any inturning of tissue (but the situation in primitive teleosts, like trout, *may* be different), notochord, neural tube and gut appear as three rods in this ridge, and the latter two develop cavities. Eyes develop at the sides of the brain and gill pouches form at the anterior end of the gut; the heart appears on the yolk, ventral to the head (Fig. 27 (b)). We, therefore, have the odd phenomenon of a little fish formed on top of a ball of yolk, with a blastopore which is still open (Fig. 27 (c), (d), and Plate IX); compare Fig. 9 and 42.

As the cellular blastodisc spreads over the yolk, cells leave its boundary and penetrate deeper into the substance of the yolk as the **periblast.** This usually occurs as a peripheral ring around the extending blastoderm which is relatively restricted to the surface; frequently, too, the periblast extends deep into the yolk *under* the embryo as a thread into the centre of the yolk (Fig. 27). As the gut tube is formed from the original undersurface of the blastodisc, it will be seen that the yolk of the teleost fish must effectively come to lie in the *coelom* (peritoneum) of the little fish, where the periblast cells continue to invade and digest it. A few authors have suggested that some at least of the periblast cells are not in fact cells of the embryo proper, but derive from supernumerary sperm nuclei resulting from the polyspermy which is so common in these animals. In any event, periblast cells do not seem to contribute to the organs of the adult but are a purely embryonic adaptation to meet embryonic needs; they are an example of caenogenesis (see p. 147).

The type of development described above is shown by all teleosts, even the most primitive of them and is quite different from that of all other vertebrates. The relationship of cytoplasm to yolk, the kind of gastrulation (with no inturning of notochord, no open neural tube, peculiar formation of gut and the yolk itself in the coelomic cavity) all must have been developed in the early history of the common teleost ancestral stock.

The selachian fishes, the dogfish, rays and sharks, have a kind of development which resembles the reptiles and birds much more than the teleosts. This is in accord with the prevailing belief that their ancestry remained like the common vertebrate stock. So most zoologists see the resemblance of their embryology to that of the tetrapods as a case of retention of the primitive mode of development and not as direct evidence of close phylogenetic relationship. Most of these fish have large yolky eggs greatly resembling the "yolk" of the chicken egg in size and colour. The tiny blastodisc cleaves and forms a plate of cells; the posterior margin tucks under (future notochord and gut roof) and cells pass into the space between the two layers as mesoderm, splanchnopleure against the lower layer and somatopleure on the under side of the upper layer. The edge of the blastoderm, which is now four layered (ectoderm–somatopleure–splanchnopleure–endoderm), now spreads over the yolk to enclose it *inside the endoderm;* when the process is complete, the yolk is in the gut ready for normal digestion. There is apparently no structure of cells comparable to periblast. Meanwhile, the notochord has induced neural tissue from the outer layers over it, which has gone on to form spinal cord and brain: once again, we have a little fish sitting on top of a ball of yolk with the blastopore still open, but the route by which we have achieved it is quite different from that of the teleosts. This blastopore is again represented by the advancing edge of the blastoderm in its mechanical function of engulfing the yolk, rather than its epigenetic function of changing the fate of tissue which passes over it as in the frog.

Many of the selachians are ovoviviparous or viviparous. They show different extents of dependence of the embryos on yolk or maternal fluids, etc., measured by their increase of (dry) weight while in the uterus of the mother fish.

The tail of fishes is formed in much the same way as that of the tadpole (p. 36); and the unpaired fins (dorsal, tail and anal), originally present as a continuous fold along the back and belly and continued around the tail, only grow and acquire skeletal support in the positions characteristic of the larva. The paired fins (pectoral and pelvic) appear as "lappets" on the flanks which become progressively constricted at their bases.

DEVELOPMENT OF THE CHICK

Here again the egg is telolecithal, but the situation is complicated to some extent by the addition of large quantities of albumen, the shell and other tertiary membranes. The structure of the bird egg is shown in Figs. 4 and 28. The blastoderm is visible as a circular white area on the surface of

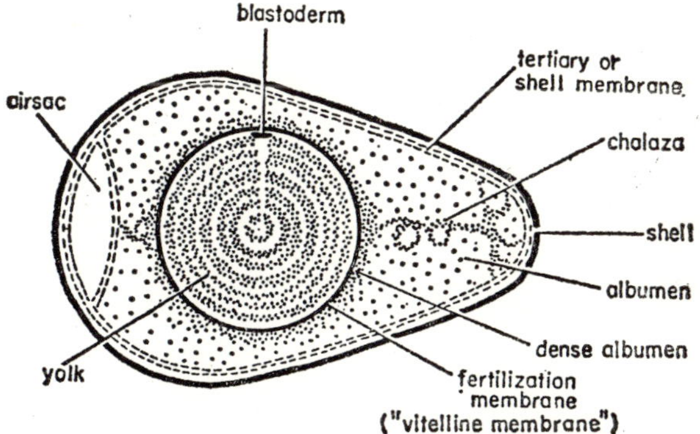

FIG. 28. *The hen's egg. Highly diagrammatic.*

yellow yolk, underneath the vitelline (or yolk) membrane, really the fertilization membrane. It has been fertilized high in the oviduct and, although many sperms have penetrated the membrane, it is virtually certain that, as in the guppy, only one unites with the egg nucleus. Development starts while the tertiary membranes are being added by the oviduct, "uterus" and shell gland. After a sequence of vertical divisions, occasional horizontal divisions occur which separate the lower layer or **hypoblast** from the upper layer or **epiblast.** The posterior end of the hypoblast probably has a more complicated origin from the epiblast. This blastula should be regarded as having nominal blastocoele between epiblast and hypoblast. This is the condition in which the egg is laid, and eggs may be stored in a cool place for some days without deterioration or further development.

When the eggs are incubated, however, there may very soon be noticed an odd appearance in the centre of the epiblast. The appearance is very similar to the wrinkle which would be produced on porridge by a nail dropped on to it. By marking points on the epiblast with vital stains, it can be seen that tissue is moving from either side toward this **primitive streak** (Fig. 29) and that the appearance of the "head" end is due to tissue moving backward in the axis of the future embryo and then rolling over and passing under itself (Fig. 30). Here, at the anterior end of the streak, notochord is being produced anterior to the streak itself, between the

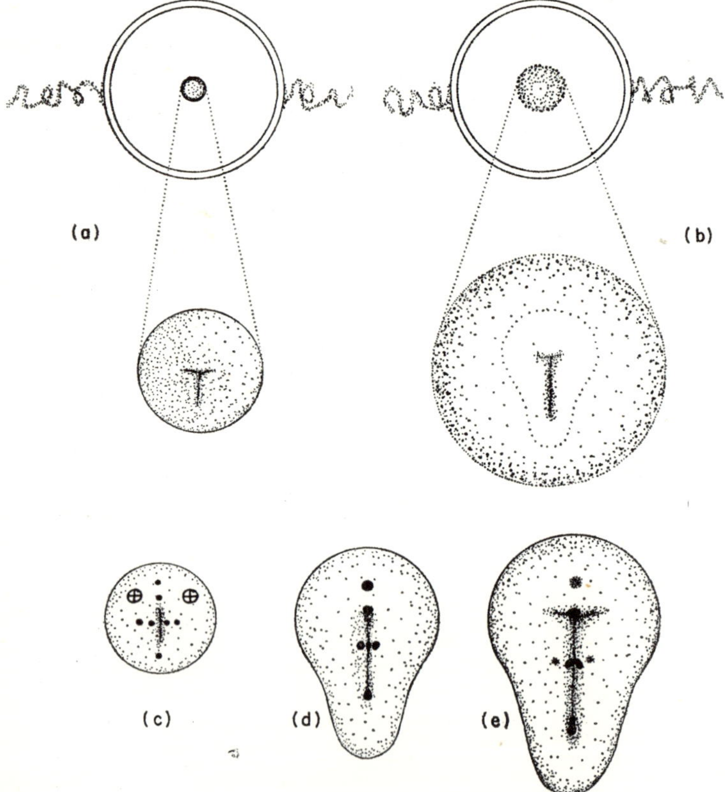

FIG. 29. *Early development of the chicken egg. (a) Yolk with blastoderm unincubated. (b) After 12 hours incubation. (c), (d), (e), stain marks (shown black) have been made around the primitive streak and their positions drawn at 30-minute intervals. On (c) the areas of heart-forming cells are marked with* ⊕

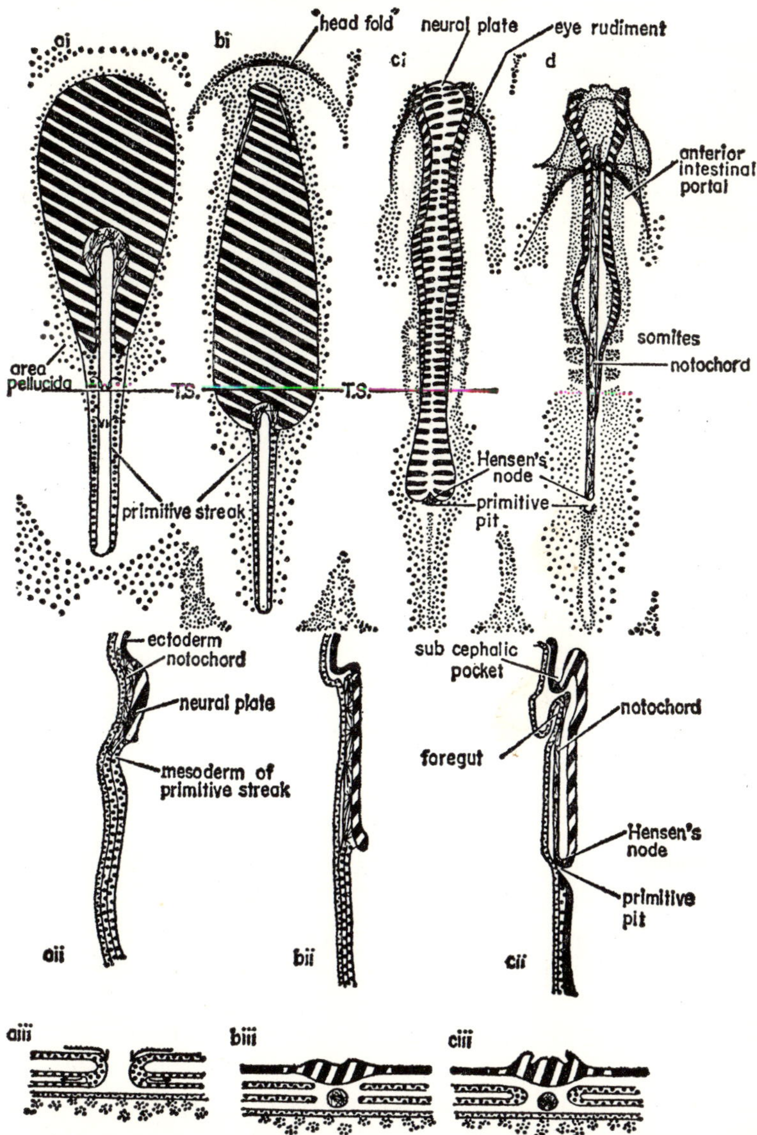

FIG. 30. *Chick fate map transformations.* (*a*) *The primitive streak stage.* (*b*) *The neural plate has formed in front of the primitive streak.* (*c*) *The neurula (compare Plate XII).* (*d*) *Stage* (*c*) *in transparency. i. In plan view. ii. In vertical longitudinal section. iii. In transverse section at level T.S. in* (*a*)*i,* (*b*)*i,* (*c*)*i.*

hypoblast and the epiblast. The hypoblast is not yet involved in the movement. Tissue rolling into the streak contributes a little to the middle part of the hypoblast; the original hypoblast is pushed laterally by this tissue, which will form the gut roof endoderm. The original hypoblast contributes only to the yolk sac (see below) and to the ventral part of the midgut.

Along the length of the streak, epiblast tissue is moving from further laterally, rolling over the lips of the streak and mostly passing into the space between epiblast and hypoblast (Plate VIIIa), where it splits into two layers, obviously splanchnopleure and somatopleure again. These two layers proceed antero-laterally on either side of the notochord and are in close proximity to the hypoblast and epiblast respectively (see Fig. 30). That epiblast tissue which finds itself above the notochord, after the latter has rolled in, becomes transformed into a neural plate, as in the frog and selachian. This neural plate rapidly transforms into a neural tube as in the frog.

Meanwhile, the gut endoderm (which probably came from epiblast through the early primitive streak) arches up in the midline much as does the gut of the frog, to form the roof of the gut of the chick embryo. It must be emphasized that this gut has no *floor* at this stage. While these events are occurring the edges of the blastoderm are extending, almost certainly with *no* rolling under of tissue, or at most very little. This covering of the yolk is not in fact complete until just before hatching and even then a pore may remain (Fig. 33).

The somatopleure which passes anteriorly by the sides of the notochord becomes organized into discrete clumps called **somites** (Fig. 31). These will form much of the dorsal musculature and other dorsal mesodermal organs of the chick and their number is added to at the posterior end as more epiblast dives into the primitive streak and is transformed into mesoderm.

The axial endoderm, which is already arched, now bulges forward under the head of the embryo, producing a pocket of **foregut.** The threshold of this pocket is called the **anterior intestinal portal** (old usage of word "portal" to mean "entrance") or **AIP** (Plate VIIIb and Fig. 30 (d)). It is within the mesoderm lying in the fold under this that the **heart** soon develops (Figs. 31 and 33). A **posterior intestinal portal (PIP)** also appears

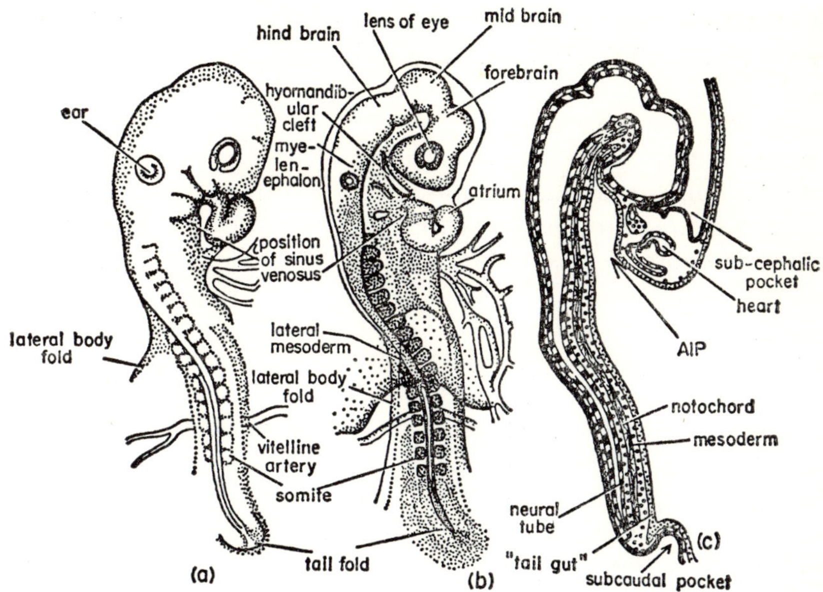

FIG. 31. *Later development of chick embryo of about 27 to 28 pairs of somites (51—56 hours of incubation). (After Nelsen 1953). (a) External view. (b) Transparent whole mount. (c) Sagittal section diagrammatic.*

as the hind end of the gut forms a pocket (Fig. 33). A tail now appears as in the fish and frog (but, of course, does not contribute much to the chick).

Having followed the development of the chick to a stage comparable to that illustrated in Fig. 9, let us compare the various ways we have described by which vertebrates reach this state. The anterior end of the primitive streak of the chick, where notochord is being produced (**Hensen's node** or the **primitive knot**), obviously corresponds to the dorsal lip of the frog. The primitive streak itself, where the more lateral mesoderm is formed by **involution,** must be comparable to the lateral and ventral lips; here then, we find a blastopore which comes into existence already closed! It is only by considering it in these terms that the early embryology of frog and bird may be compared. The teleost fish, on the other hand, develops notochord, nerve cord and other axial structures without overt tissue movements.

COMPARISONS AND MECHANISMS IN EARLY VERTEBRATE DEVELOPMENT

We have now followed the development of the eggs of fish, amphibian and bird until their attainment of the chordate morphology is obvious. There are various reasons for believing that this early part of the development of an organism is controlled to a large extent by the genes of the mother, acting via the organization of the egg cytoplasm and cortex. The genes of the zygote only begin to exert their effects after gastrulation and, especially, after the attainment of the "generalized" condition for the group of animals concerned. For the vertebrates this condition is represented by Fig. 9. It will have been observed that although the eggs of fish, frog and bird differ greatly from one another, as do the adults, all pass through this stage of the simplified vertebrate condition and the resemblance of these organisms at this stage is very striking. Equally, the embryos of the other phyla of the animal kingdom are frequently very diverse from fertilization through cleavage, but allied forms come to resemble one another just prior to organ formation; then they diverge again as development proceeds. There seems to be a "stable" **phyletic stage** in development; before this the eggs vary in the amount of yolk they carry, in their size and shape, their cleavage patterns (e.g. frog, fish and bird) and in the organization which prepares them for gastrulation. After this they again come to differ as the specific adaptations of the adult are formed.

There is now considerable evidence that up to the attainment of this phyletic stage the genetics of the mother controls development, because the organization of the oocyte is by *its* gene complement, i.e. the maternal genome; this organization results in development towards the phyletic stage *independently* of the genes in the zygote nuclei. Only after the phyletic stage has been achieved can the zygote genes exert their specific effects—only now do they find themselves in the appropriate position in the embryonic geography.

Some of the evidence for the view that zygote nuclei only begin to make their presence felt later in development derives from Moore's elegant work on nucleocytoplasmic hybrid amphibian eggs. Destruction of the egg nucleus in various ways, followed by fertilization by sperm of different species, or the transplantation of nuclei into enucleated eggs, allowed him to investigate the effects of nucleus and cytoplasm separately. In the latter case, the blastula and early gastrula resemble the species which provided the egg cytoplasm; later stages resemble the species donating the nucleus, if they survive. A critical period occurs during and immediately after gastrulation, at which time "incompatibility" between cytoplasm and nucleus may lead to malformation and death. Some frog hybrids die for this reason; the zygote nucleus is incompatible with the egg cytoplasm. In the case of the reciprocal cross, where the other species contributes the egg cytoplasm, embryos may develop normally. Presumably species may differ *either* in their nuclear or their cytoplasmic character, or in both. These amphibian embryos did not attain the phyletic stage when incompatibility occurred (presumably this is represented by the "neurula", on Fig. 9, in vertebrates); but they developed far enough to show clearly that this phase of development was under the control of the cytoplasm species. Those cases where development was completed showed more and more characters of the nuclear donor appearing as organ-formation occurred. The genes of the zygote nucleus come into play progressively in different parts of the embryo, and different parts of the common gene set are expressed in neural tissue, liver, muscle or skin (see p. 137).

It is apparent from the description of gastrulation in the frog and chick that the first event involves dorsal lip material. It has been experimentally demonstrated that the dorsal lip of the frog, or the comparable structures in fish and chick will, if transplanted under blastular tissue, cause the production of a new set of axial embryonic structures (notably notochord and nervous system). These are produced by the *local blastular tissue*, which might under other circumstances have remained as skin. These experiments demonstrate the epigenetic action of the **primary organizer**; and it is necessary for further description that we acquire a terminology. In fact, tissue from the primary organizer will not induce the production of axial structures in tissue which has *already* been under the influence of such an organizer and which is already **determined** in its **fate** (for example later neural plate). Some tissues are therefore **competent** to respond to the **cue** from this organizer; others are not. Blastular cells are; neural plate is not. There can be little doubt that this cue, which results in induction, is primarily chemical. Killed organizer tissue, or indeed substances extracted from such tissue, or even certain other biologically active compounds, may **evoke** the response in competent tissues. The chemical stimulation is called **evocation** and the chemical substances the **chemical evocators.**

The simple ideas of one evocator per tissue have now been replaced by a variety of models. These involve gradients and other spatial chemical variations and the varying responses of cells to these cues. The cues give the cells **"positional information"** to which the cells respond according to their competence. This is, in turn, determined by their original cytoplasm (all nuclei have the same genes) with its maternal m-RNA programme, as well as by the history of the cells concerned.

That one choice which one *knows* is within the power of a tissue to become, because it ordinarily does become it, is called its **potency.** The potency of neural plate is to form spinal cord and brain. The **competence,** on the other hand, is a *whole range* of possibilities open to a tissue in various experimental situations; given different cues, perhaps corresponding to the information which it would receive in a different position in the normal embryo (different positional information), the same kind of cells may prove competent to produce a wide range of tissues and structures.

We see, therefore, that the result of any interaction depends on at least three features. Firstly, the tissues involved must be in certain spatial relationships to one another. Secondly, the tissue which is to receive the cue must be competent to respond. Thirdly, the cue which is given by the inducing tissue, the evocator, must be in terms which the reacting tissue "understands". A change in pH may induce a tissue culture of anterior ectoderm of the chick to become neural, but will not affect cultures of posterior ectoderm.

The reaction to evocator may be a graded response, but very much more often is a kind of **threshold effect.** That is to say, until a certain concentration of evocator is reached there is no response, but above this concentration (the threshold) the response is complete. There may be two or three different responses possible by a competent tissue to one evocator. As an example, there has been much discussion as to the difference in the interactions which result in the production of brain or spinal cord. It seems that in some animals (e.g. the frog) the whole neural tube is evoked by one substance. This is produced copiously at the anterior end in the **prechordal plate** (the very front end of the notochord), but its concentration diminishes as one progresses posteriorly. The line of demarcation of brain and spinal cord is evidence of two successive thresholds in the reacting tissue. In other forms perhaps for example, the chick, two substances seem to be involved whose concentration gradients are inverse; "brain evocator" has a high concentration anteriorly and "spinal evocator" has a high concentration posteriorly. Note that there the tissue which reacts must have a different *kind* of competence for each evocator. Note also that around the neural plate of all vertebrates is an area which has been under the neuralizing influence of these evocators, but has not responded by becoming neural plate. Perhaps this area marks the limit of adequate lateral diffusion of the neural evocator; the neural threshold is not *quite* achieved. The cells of this region will come to lie in the neural crest as the neural plate sinks and they will form one of the most important cell populations in vertebrate development (see p. 112).

After the primary organization has resulted in the appearance of the **primary tissues** (notochord, prechordal plate, somites, neural tissue, and often gut and skin), each of these in turn organizes neighbouring tissues so that still greater complexity is achieved. Organizing reactions still go on, their complexity often increasing, until the whole embryo with all of its functioning organ systems results. There is, therefore, a **hierarchy of organizers** which commences with the primary organizer, e.g. dorsal lip of the frog gastrula, and continues with secondary organizers like the prechordal plate and notochord. We must wait to consider tertiary and quaternary organizers until we discuss the development of some vertebrate organ systems (pp. 92 et seq.) from the generalized condition in which we have left them (Fig. 9).

In many organisms, most notably the insects, development seems to occur autonomously in various **compartments.** Each compartment (a body segment, or a product of one imaginal disc) produces a variety of tissues from very few clones of cells, perhaps only one clone. This means that muscle, skin and nerve cells in one segment are more closely related to *each other* (by cell lineage) than are, for example, the muscle cells of adjacent segments, or the epidermal cells covering adjacent segments. In other words, "skin", "muscle" and "nervous system" in insects are of multiple derivation and the same tissues up and down the animal are related less closely than the nervous, muscle and skin cells in one segment are to each other. It is difficult to imagine similar compartments in vertebrate development, but the proponents of this model do make claims for its universality.

THE SEQUENCE OF DEVELOPMENTAL EVENTS

The phenomenon of **heterochrony,** or difference of timing, must be introduced in order that we may consider and compare the development of the organ systems of various animals. It is by no means an infallible rule that organ systems develop in the same order in different animals. For example it is usually the case that embryos from telolecithal eggs need and produce a blood vascular system very early in development, as compared with their homolecithal relatives. This is obviously a specialization to enable utilization of the yolk (which is *outside* the cells) to proceed. When we come to the development of various mammals, heterochrony will be seen to be extreme in this group; this is also an adaption to enable nutrients to be obtained for the embryo from outside; in this case, however, it is not from the yolk, but directly from the mother.

Let us consider for a moment the development of the ear of vertebrates (see also pp. 123–124). No one could deny that the utriculus and sacculus are homologous structures in all vertebrates. However, the mode of their formation varies considerably (Fig. 32). They are derived from the ectoderm (skin) on either side of the hindbrain as the auditory placodes

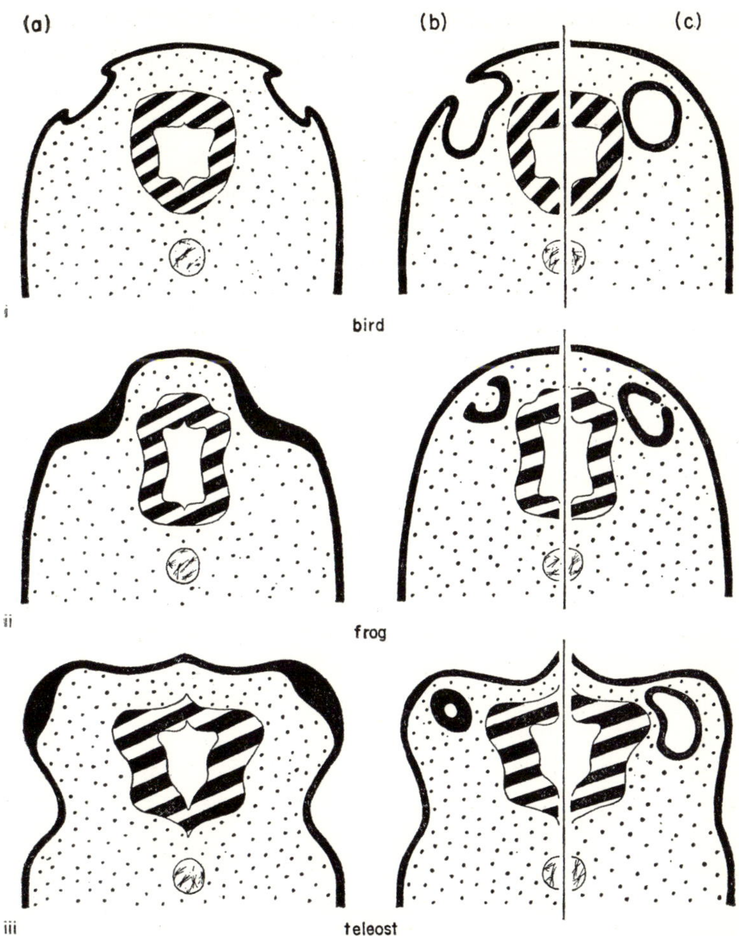

FIG. 32. *Development of the auditory placode of vertebrates.* (*a*) *i, ii, iii show the dorsal half of transverse sections through the posterior part of the head of a chick, a frog and a teleost at an early stage.* (*b*) *and* (*c*) *show later stages of placode formation in these animals, as left and right sides.*
In bird *the placode rolls up, drops in and then separates.*
In frog *the placode drops in, rolls up and then separates.*
In teleost *it "rolls up", separates and then drops in.*

(a **placode** is an ectodermal thickening which will later drop in to form a sense organ). In some forms (birds and many mammals), the placode bulges in and then drops into the underlying mesoderm as a vesicle which contains, as it were, part of the "outside world". In other cases (some amphibians), the placode drops in as a plate, which rounds up enclosing some of the mesoderm which then dies leaving a cavity. In still other cases (many teleost fishes), the placode thickens *in situ*, drops in, and then a space appears within it. As we agree that the structures produced are homologous, the several modes of production must be variations on one theme. It can be seen that by varying the *relative timing* of the dropping in, the curling up and the thickening of the layer, the three situations above can be explained in the same terms; only the sequence differs (see caption to Fig. 32).

Such cases are widely distributed in animal embryology, for example, the blastopores of frog and bird. Such differences in timing may, if no later compensation occurs, come to represent important differences between the adult animals. They will obviously affect the epigenetic relationships, because they will affect the "rendezvous" of the tissues concerned. Those embryos which develop in an environment whose temperature varies must have a variety of compensatory devices to ensure that the epigenetic relationships which depend on the meeting of two tissues will still occur successfully. This imposes an upper and a lower temperature limit on animals, outside which viable development cannot occur. This viable range is often smaller for eggs and embryos than for adults, which, anyway, are better equipped to find and use more constant environmental conditions, if these are available. Nevertheless, the eggs of *Rana temporaria* will develop over a range from 5 °C to 25 °C and in the British springtime a change of temperature of 12 °C between day and night is not unusual; normal tadpoles still appear. There is, however, considerable evidence that the compensatory mechanisms can only retain the necessary **associations** within a given range of temperature and **dissociation** of embryonic processes occurs outside this range; this causes failure of development. One evolutionary route away from the risk of dissociation is to maintain the embryos within the parent organism; even if the parent is not a homeotherm, its nervous system will act as a fairly efficient thermostat, as it keeps its own body at an equable temperature. For example, the developing eggs within a female guppy will meet much less extreme temperatures than those of a frog, as the female seeks water at her optimum temperature and is readily able to move to it.

There are also evolutionary consequences of the association and dissociation of developmental events. Not only is the sequence of developmental events in closely related organisms often different (ABCDEFG compared with ABDECGF), but one organism may produce a structure not comparable to anything in the other, or one may omit characteristic structures which the other possesses. For example, the polychaete worms develop via a ciliated trochophore larva, while the earthworm has omitted the locomotory larval stage (p. 47); tadpoles of frogs usually have horny sucking lips and no "balancers", while tadpoles of newts have "balancers" and no horny lips; embryos of viviparous fishes and reptiles have a placenta for nutrient exchange with the maternal tissues, embryos of oviparous forms do not. These are all embryonic or larval structures developed to fill a need of the organism *at that age;* they do not perform any function in the adult and are usually lost at metamorphosis or birth. Such interpolations into the developmental sequence are called **caenogenetic** and are very common. Perhaps, the largest single category of them is the embryonic membranes. The embryonic membranes of mammals also show extreme heterochrony in that they are usually formed from the egg before the embryo itself appears.

EMBRYONIC MEMBRANES

In general, those animals whose eggs develop rapidly into free-swimming larvae are sea-water forms which lay very many eggs (see p. 14). Other forms, notably the insects, the cephalopod molluscs and land forms in general, lay fewer eggs and provide them with sufficient reserves to enable them to hatch as competent organisms which are more or less capable of fending for themselves. Embryonic development of such forms must allow for the utilization of these reserves, which are usually in the form of yolk. A simple solution to this problem occurs in many land and fresh-water snails; here, the egg itself is homolecithal, like those of its sea-water ancestors, but lies in a very nutrient fluid in the egg capsule (Plate IIIb). More usually, the egg cell itself contains the yolk. In fishes and birds, with telolecithal eggs, the edge of the growing blastodisc engulfs the yolk and becomes the **yolk sac.** As the yolk is absorbed, so this retracts until finally it forms the ventral part of the body of the young animal. Other **embryonic membranes** are the **amniotic sac** (whose walls are the **amnion** and the **chorion**) and the **allantois.**

The amniotic sac is found in the fishes, reptiles, birds and mammals, and a comparable structure is found in many invertebrates (Fig. 23). Around the area where the embryo is forming on the blastoderm, the outer layers of the blastoderm elevate and form a fold (like a ruck in a table-cloth), which comes to enclose the embryo by a process resembling the closure of a string purse (Fig. 33). A pore may remain, the **amniotic pore.** *Two* membranes now enclose the embryo, an outer chorion and an inner amnion. In vertebrates, both are ectodermal and lined with vascular somatopleure. The chorionic (and yolk-sac blood) vessels are the main organs of respiration of the guppy embryo; and it is probable that waste products also diffuse into the mother's blood from them.

At about 50 hours of incubation, the chick embryo has a well-established blood vascular system and is comparable with Fig. 31. The amniotic fold appeared, anteriorly at first, and then tissue was pinched up more laterally and finally more posteriorly (Fig. 33 (a) and (b)). At 72 hours, the amniotic fold appears as a veil drawn over the embryo (Plate VIIIe) and by 80 hours only a pore remains connecting the amniotic cavity with the albumen (white) of the egg (Fig. 33). The fertilization membrane has ruptured at about this time.

Meanwhile, the foregut and hindgut of the embryo have been increasing in length as the embryo itself grows; successive diagrams of development drawn to the same size (e.g. Fig. 33) make it appear as if the AIP and PIP are approaching each other—in fact, the distance between them remains the same, but the embryo grows fore and aft of them; it is, however, convenient to speak of "the AIP and PIP approaching each

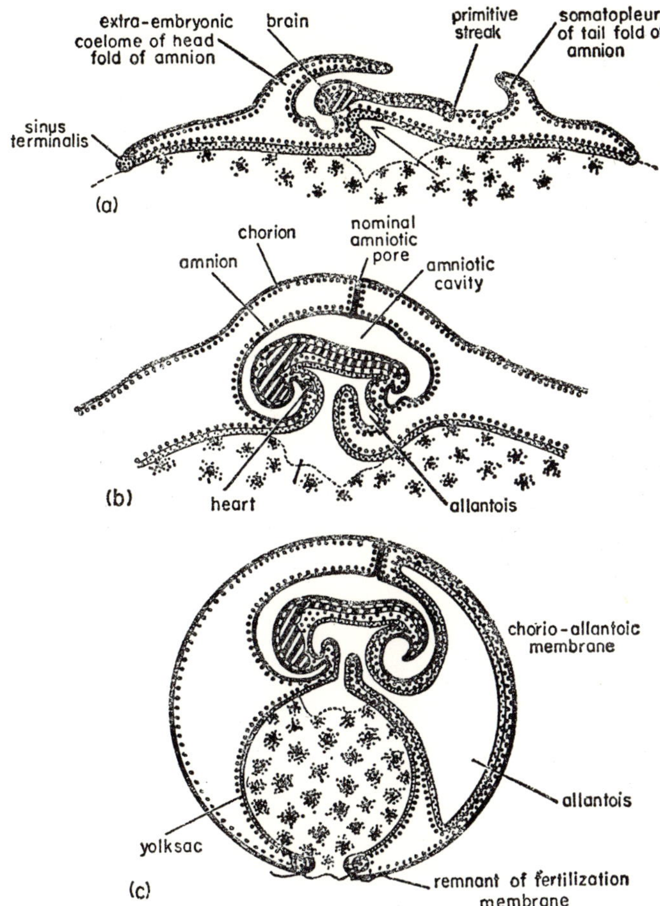

FIG. 33. *The embryonic membranes of the chick, in vertical longitudinal section. (a) The amniotic folds are appearing and the foregut is lengthening. The AIP is denoted by the arrow; about 50 hours incubation. (b) The amniotic folds have met and the allantoic bud is dropping into extra-embryonic coelom; about 85 hours incubation. (c) The allantoic bud has expanded and its splanchnopleure has fused with somatopleure of the chorion. The margin of overgrowth (denoted by the sinus terminalis) has left only a small pore into the yolk sac, around which the fertilization membrane may persist.*

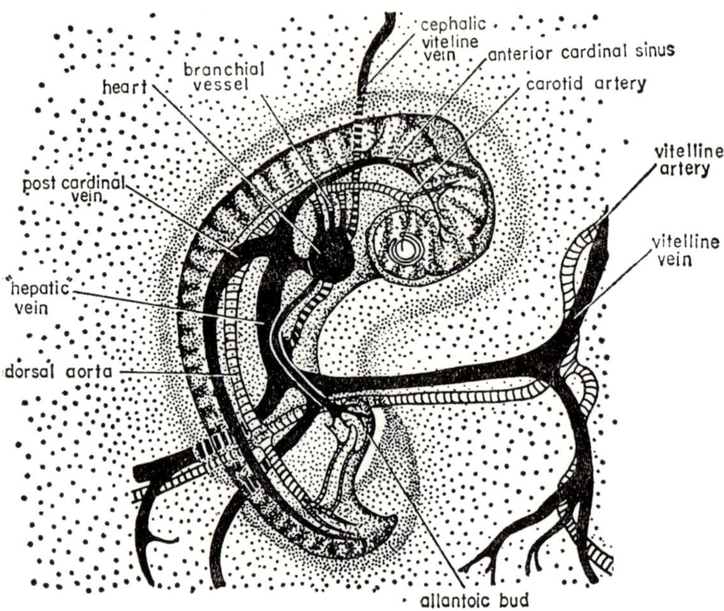

FIG. 34. *An 80-hour chick, seen in transparency, with blood vessels labelled.*

other" until the gut is connected to the yolk-sac only by a narrow stalk within the **umbilicus.**

Let us now consider the structure of the membranes at this stage (Fig. 33 (c)). Up to the **margin of overgrowth** the outer membrane is the ectodermal **chorion,** lined by vascular somatopleure; an extension of the embryonic coelom separates this from the somatopleure on the *outside* of the amnion, which is, of course, also ectodermal. Within this is the amniotic cavity, in which is suspended the embryo on its **yolk-sac (umbilical) stalk.** Within this stalk is the endodermal tube (representing the AIP and the PIP which have "met") connecting the gut of the embryo with the yolk. On its *outside*, this is covered with very vascular splanchnopleure, whose vessels transport yolk products, via the vitelline veins (p. 95) in the stalk, to the embryonic liver (p. 103).

Initially, when the embryo is very small relative to the size of the whole egg, its principal nitrogenous excretory product is ammonia. However, the bird's egg is a closed system (a **cleidoic egg**) and ammonia is very toxic. The embryo soon forms much urea and then, as even urea is toxic in quantity, turns over to uric acid as its main nitrogenous excretory product. This is insoluble, and the crystals are stored in the third embryonic membrane, the **allantois.** The allantois originates as a ventral diverticulum of the hindgut into the coelom and may first be seen as a "bubble" at about 80 hours of incubation (see Fig. 34). The hindgut is, of course, covered with splanchnopleure and the allantoic bud remains invested in this vascular (mesodermal) coat as it elongates. It must be emphasized that the allantois pushes out into the *coelom.* As it increases in volume, it soon fills the coelom ventral to the hindgut and bulges under the PIP into the coelom of the yolk-sac stalk (Fig. 33 (b)). Its distal end now lies in the **extra-embryonic coelom** and as the allantois expands still further its splanchnopleure approaches and finally fuses with the somatopleure on the inside of the chorion, forming the **chorio-allantoic membrane** (Fig. 33 (c)). Both layers of mesoderm become extremely vascular and this area becomes the main respiratory organ of the forming chick; it closely underlies the tertiary membranes just within the egg-shell.

Reptiles, birds and mammals all possess very similar systems of embryonic membranes and in the mammal these contribute to the embryonic part of the **placenta.** In those reptiles in which the young are born alive, for example the adder, close contact is established between the chorionic blood supply of the embryo and the mother's circulation, although it is rare for the fertilization or tertiary membranes to break down. Frequently, the blood vessels of the yolk-sac or the allantois also enter into this relationship, and then the organ is called a **yolk-sac placenta** or a **chorio-allantoic placenta.** Yolk-sac placentae are found commonly among the fishes, especially the selachians (see p. 62) and nearly all the marsupials have a yolk-sac (plus chorion) placenta.

THE DEVELOPMENT OF MAMMALS

Only now have we laid a proper foundation for a consideration of the early embryology of the major mammal groups. However, before considering them, passing mention must be made of the egg-laying mammals, the monotremes.

In this group development resembles that of reptiles and birds. The egg is telolecithal and the yolk comes to be invested in a yolk-sac, just as in the chick; the formation of amnion, chorion and allantois also follows a similar path. Indeed our illustrations of chick development will also apply to the *Platypus*.

The marsupials and eutherian mammals all have alecithal eggs; all but the very earliest nutritional requirements must be drawn from the mother's circulation and because of this extreme heterochrony has resulted in the appearance of some organ systems, notably the embryonic membranes, before the embryo itself has been established. The eggs have no true vitelline membrane and on discharge from the **Graafian follicles** of the ovary are surrounded by the cells of the **cumulus oophorus** and have a thick **zona pellucida** (Fig. 35). The sperms pass between these cells, which

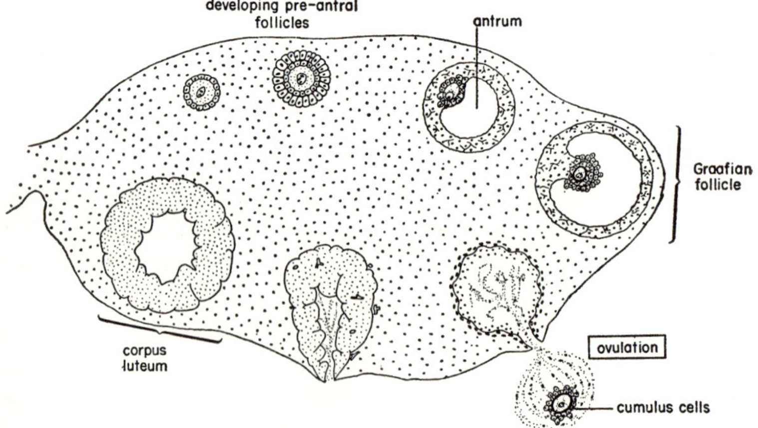

FIG. 35. *The ovary of the mammal. showing the progression of a follicle through ovulation to become a corpus luteum.*

then fall off. The egg membrane elevates and becomes a kind of fertilization membrane. After fusion of the gamete nuclei, cleavage divisions commence. These result in a morula of 8–16 loosely aggregated cells, of which one or two may already have become specialized because of enclosure within other cells. This specialization occurs only after **compaction** of the morula, when the cells adhere closely to each other, giving a smooth outline to the forming blastocyst. In some armadillos, each of 4 blastomeres produces one of a family of quadruplets; each blastomere contributes to a large communal blastocyst in which four embryos develop, sharing a common placenta.

The egg as it divides has been rolling down the Fallopian tube and usually falls into the uterine lumen after a cavity has appeared among the completed cells. It is now known as a **blastocyst** (Fig. 36). In some marsupials, the blastomeres separate and then press against the inside of the fertilization membrane, reforming the spherical blastocyst; this emphasizes, as do their subsequent histories, that the blastocyst and blastula cannot be compared. Almost all of the cells forming the wall of the blastocyst compose the **trophoblast.** Only a small proportion of their number will form the embryo and other membranes. Cells of this small area bulge into the cavity of the blastocyst forming the **inner cell mass.** The wall of the blastocyst, the trophoblast, begins to erode the uterine epithelium and its cells fuse into a syncytium, the **syncytiotrophoblast**, then push out processes which invade the deeper tissues. These cells do not divide, but the chromosomes in their nuclei replicate and they become extremely polyploid. There are, in fact, two populations of cells on the outside of the blastocyst, the **mural trophoblast** whose cells form relationships with the mother, and the **apical trophoblast** lying above the inner cell mass, whose cells divide and add to the area of the trophoblast during development. Embryos which lack an inner cell mass fail to produce apical trophoblast, so it is clear that the inner cell mass maintains division in these trophoblast cells. The question of which cells of the morula become trophoblast and which become inner cell mass has been investigated experimentally by a number of very elegant techniques. It now seems clear that any cell of the morula of the mouse, when injected into a slightly older blastocyst, can contribute to either trophoblast or inner cell mass.

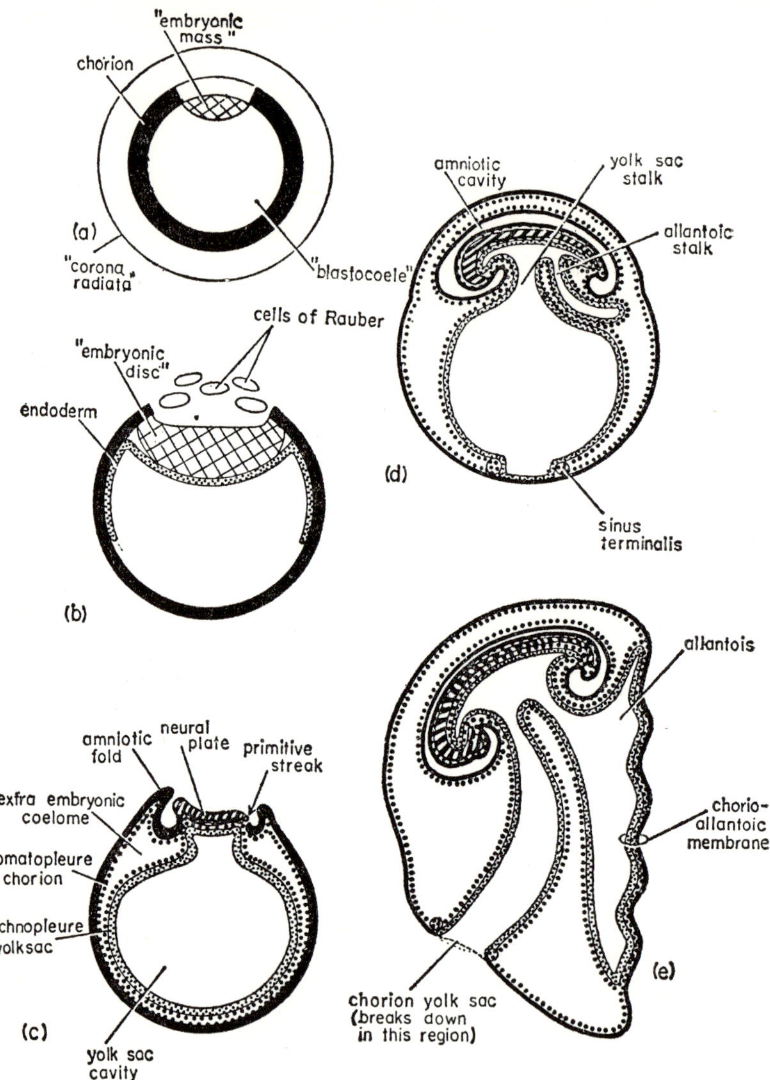

FIG. 36. *Development of the rabbit.* (*a*) *The early blastocyst.* (*b*) *Exposure of the "embryonic disc". Endoderm is being delaminated from its undersurface and is lining the yolk sac.* (*c*) *A primitive streak on the embryonic disc has formed the embryo and the extra-embryonic mesoderm. The yolk sac is complete.* (*d*) *The amnion has closed and the allantois has appeared.* (*e*) *The yolk sac floor has broken down within the ring of the sinus terminalis, allowing antibodies from the mother access to the embryo.*

Further, any cell of inner cell mass or apical trophoblast from an early blastocyst may become any part of a later blastocyst—only the mural trophoblast is already determined. Little is known of those cell interactions within the inner cell mass which determine the future fate of the different cells, but it seems very clear, in contrast to nearly all other animals, that the mammal embryo does not depend upon the architecture of its egg for the organization of its parts.

Whether there is a developmental programme dependent upon m-RNA received via the oocyte is also not completely clear. In other organisms, like the Amphibia, the production of substances or characters from the paternal genome has been used to distinguish effects of the zygote nucleus from those of the maternal programme. For example, proteins produced during cleavage in Amphibia are always characteristic of genes of the maternal line; only after gastrulation can specific proteins (usually enzymes) of the paternal line be found in the embryo. In mammals, however, the paternal genome produces some of its characteristic enzymes as early as the 4-cell stage (possibly only in mural trophoblast when cleavage has been delayed). However, the mammal embryo is usually 2 or even 3 days old by this time; when the fish or amphibian developing at 37 °C is 72 hours old, it has probably finished gastrulating. In any event, its zygote genome is well into its stride—it may simply be that mammals cleave very slowly, but that the rest of the developmental programme proceeds at a "normal" rate.

Such slow cleavage and the lack of egg organization seems to be common to all mammals before the blastocyst stage. Differences in subsequent development relate, mostly, to different degrees of heterochrony of the embryonic membranes. We will commence with development of the rabbit, where heterochrony is not extreme, and then consider other forms.

THE RABBIT

The blastocyst of the rabbit is an ovoid when it attaches to the uterine wall, with the inner cell mass adherent to its wall midway along its length. The trophoblast cells now flatten considerably and the cavity layer of the inner cell mass towards the blastocyst cavity becomes recognizably different from the rest. Its cells divide and migrate outward onto the inner aspect of the trophoblast (Fig. 36 (b)). These cells will form the gut lining, yolk-sac and allantois and hence are the endoderm. Meanwhile, those outer cells overlying the inner cell mass (the **cells of Rauber**) break apart and are lost, exposing the outer aspect of the inner cell mass, the embryonic disc, to the uterine cavity (Fig. 36 (b)). A primitive streak now appears on the **embryonic disc** epiblast and the original middle of the disc dives into this and emerges *between* the disc and its underlying endoderm as two layers, one against the epiblast (obviously somatopleure) the other against the endoderm (splanchnopleure). Over one end of the streak (**Hensen's node**), presumptive notochord passes, as in the chick, and transforms into notochord anterior to the streak, which induces the overlying epiblast tissue of the disc to become neural. Foregut, AIP and heart appear as in the chick. The amniotic folds appear and roof over the embryo to complete the chorion. The hindgut and PIP appear and the allantois; this latter rapidly enlarges and bulges under the PIP into the extraembryonic coelom, where its splanchopleure soon joins the somatopleure of that area of trophoblast to form a **chorio-allantoic placenta** which enters into very close relationship with the mother's circulation.

The somatopleure and splanchnopleure never reach the furthest point from the primitive streak; here, endoderm and ectoderm trophoblast or (chorion) are in contact for a while, but then the area breaks down. This leaves an opening from the uterine cavity into the yolk-sac and so into the gut of the embryo (now called a **foetus**). It is by this route, in the rabbit, that antibodies from the mother are acquired by the embryo and confer **passive immunity** upon it until it can make its own antibodies. The embryos of some other mammals, like mouse and human, receive antibodies via the placenta itself, while others (cattle) receive them with the milk and absorb them through the gut (without destroying them by digestion).

EARLY DEVELOPMENT OF OTHER MAMMALS

In the marsupials, the allantois rarely touches the chorion and the placenta is a yolk-sac (+chorion of course) placenta. The embryonic disc apparently develops directly from the outer layer of the trophoblast overlying the inner cell mass. (It will be recalled that in the rabbit these cells, the cells of Rauber, break up and expose the outer surface of the inner cell mass.) Figure 37 shows the early development of mouse, marsupial and primate (Rhesus monkey) for comparison with the rabbit.

In the mouse, not only the chorion and yolk-sac but also the amnion is **precocious** (i.e. appears before the embryo proper). A cavity appears in the inner cell mass as the endoderm cells begin to migrate from it and in the floor of this cavity appears the primitive streak. The cavity, therefore, is the amniotic cavity. As somatopleure from the streak passes outward, it remains against the wall of this cavity and so provides the amnion, as well as the chorion, with mesoderm. This form of development, where the embryo is never in contact with "outside world" as represented by the uterine lumen, was called development with **entypy of the germ.** It is probable that some of the extraembryonic endoderm in this form is delaminated directly from the trophoblast wall instead of deriving from the inner cell mass (see Plate VIIb).

In the primates entypy of the germ also occurs; much of the extraembryonic endoderm (yolk-sac and allantois) and also much of its splanchnopleure and corresponding somatopleure arise directly from the blastocyst wall, the primitive streak only providing mesoderm for the embryo itself and for the yolk-sac and allantoic stalks. The anterior end of the primitive streak in primates is also interesting in that it resembles the **notochordal pit** of reptiles rather than the Hensen's node of birds

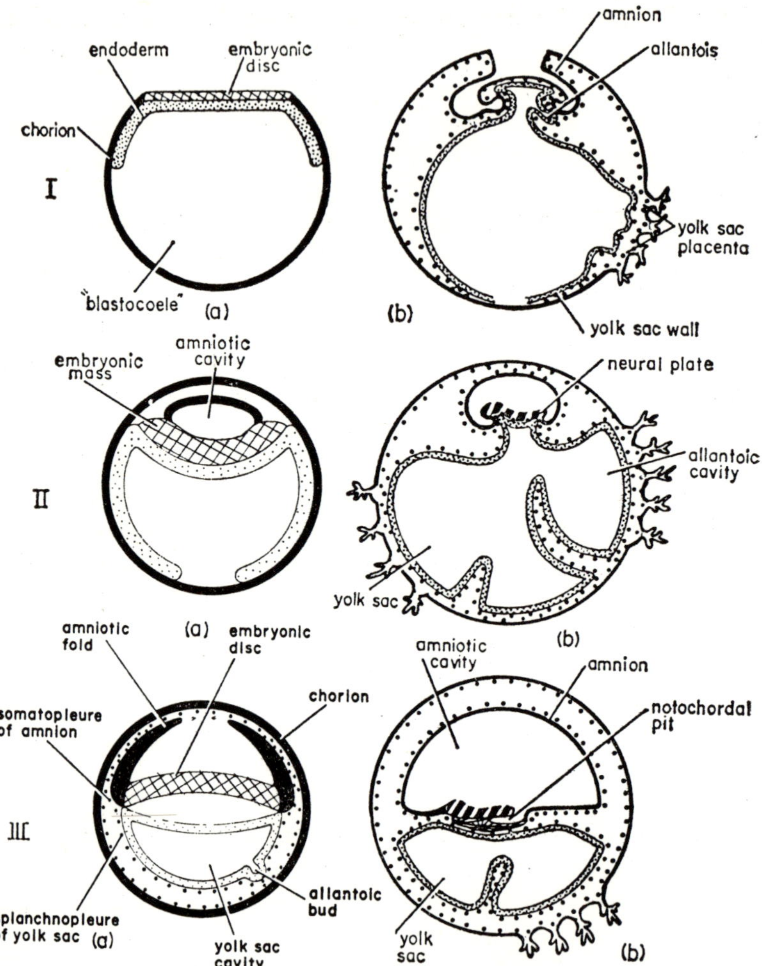

FIG. 37. *Heterochrony in the embryonic membranes of different mammals, highly diagrammatic. I. Marsupial. II. Mouse. III. Primate.* (*a*) *and* (*b*) *are two stages,* (*a*) *just* before *the embryo appears and* (*b*) *after establishment of the placenta.*

there is a real dorsal lip overhanging a tube which connects to the space below the blastoderm, the future gut. This is only possible because the endoderm is *not* complete in the midline, where notochord will form. This is exceedingly reminiscent of the situation in the frog, where the notochord may, for a short time, form the midline of the roof of the gut.

EARLY HUMAN DEVELOPMENT

The development of the human embryo can only be understood in terms of extreme heterochrony; all the embryonic membranes are so precocious that all, along with their mesodermal components, are virtually complete before the embryo is really formed. Reference to Fig. 38 and its legend, and comparison with Fig. 37, should make this clear. Reference should be made to texts of human embryology for the details and to Table 2 for the human chronology. This Table outlines the stages of development of the human at various times during pregnancy; we include it so that those who read this book may be better informed in debate on the pros and cons of abortion.

THE MAMMALIAN EMBRYO AS A HOMOGRAFT

In some cases, like the guppy, slow-worm and sharks, viviparity does not involve intimate contact between the mother's tissue and those of the embryo since there is a membrane between them. So the mother would not be expected to react against the embryonic tissues as if they were a graft from another organism and reject them. On the other hand, there must be diffusion problems for the embryo in these cases; nutrients, gases and excretory products cannot pass between mother and embryo with the same ease as in the mammals.

TABLE II. The development of the human embryo from fertilization to parturition

*Age	Length (crown/rump) (mm)	Weight (g)	Appearance and state of development
1 week	"0.1"	0.01	Flat embryonic disc
2 weeks	"0.2"	0.02	Primitive streak evident on embryonic disc; neural groove appears
3 weeks	2	0.08	Neural tube formation begins; up to 16 somites present; first branchial arches appear; body shape cylindrical, constricting off from the yolk-sac
4 weeks	5	0.15	All somites formed (40); branchial arches complete; yolk-sac stalk slender; flexed, tubular heart present; limb buds apparent; body in a C-shaped curve
5 weeks	8	0.25	Heart, liver and mesonephros all prominent, nasal pits present; tail development obvious; umbilical cord becomes more organized
6 weeks	12	0.45	Jaws become evident, though upper jaw components still separate; head enlarges and neck flexure develops; external ears become obvious; limbs now recognizable
7 weeks	17	0.7	Branchial arches disappear; facial features forming; digits beginning to form; back begins to straighten and lose its previous C-shape; tail begins to regress
8 weeks	23	1	Facial features well formed though not yet looking very "human" — nose very flat, eyes very widely separated; digits well formed; body rounding up as gut grows; now called a *foetus*
10 weeks	40	5	Head now erect; nails appearing on digits; limbs fully formed externally, though the internal structures (bones and muscles) are still not organized
12 weeks	56	14	Face beginning to take on human proportions, nose gains a bridge; head still very large compared to body; sex of foetus now visible externally
16 weeks	112	105	Face fully "human"; body now outgrowing head; hair appears; muscles become active
20 weeks 5 months	160	310	Enamel and dentine deposition on (unerupted) milk teeth; nasal bridge and ear ossicles become bony; nail growth continues; epidermis cornifies; mammary primordia begin budding

6 months	203	640	Body lean but proportions begin to take on those of the baby; permanent teeth primordia form; nostrils open
7 months	242	1080	Eyelids open, lens tunic now very vascular and retina sensitive to light; cerebral cortex of brain now convoluting rapidly

At the time of writing (October 1980) a pregnancy may be legally terminated up to this stage in embryonic development

8 months	277	1670	Fat beginning to collect in body, smoothing out wrinkles in skin; buds in mammary gland become hollow; testes descend into scrotal sacs
9—10 months Full term (average 38 wks)	350	3300	Body smooths out and becomes rounded; nails reach finger tips; pulmonary blood vessels near completion (about two-thirds complete at birth) while some foetal blood passages discontinue; kidney tubule formation ceases at birth; myelinization in the brain begins

* Fertilization is usually assumed to have occurred two weeks prior to the first missed menstrual period (or timed from two weeks *after* the start of the last observed period).

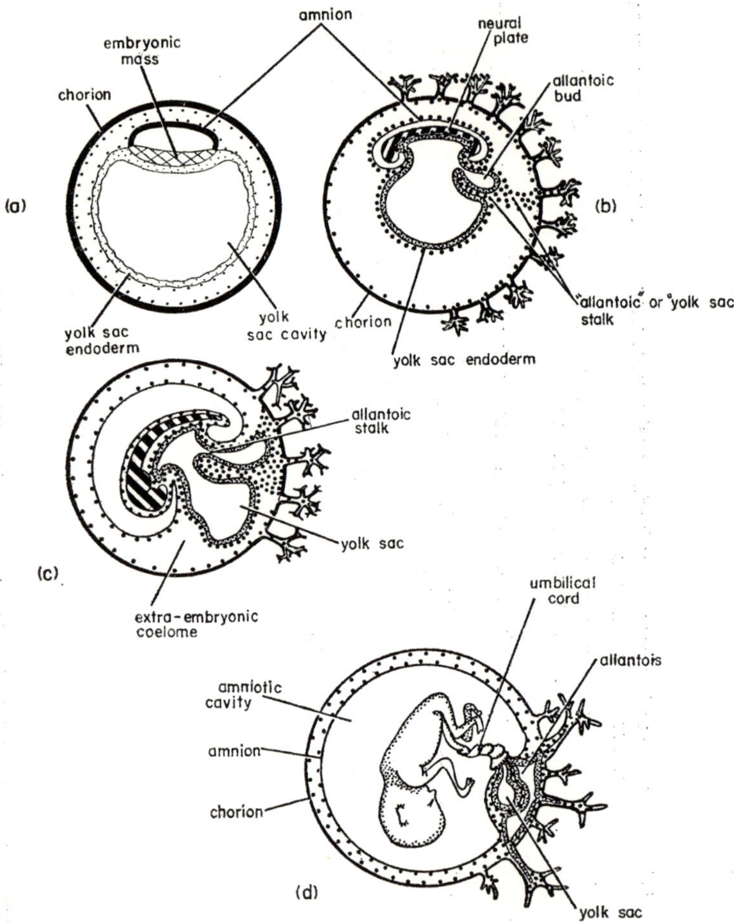

FIG. 38. *Development of the human. Note the relatively enormous amniotic cavity and the tiny yolk sac in* (*d*). (*a*) *About 12 days.* (*b*) *About 16 days.* (*c*) *About 28 days.* (*d*) *10—12 weeks.* (*after Nelsen 1953 and others.*)

In the mammals, placental contact between maternal and embryonic tissues is (at least) directly between syncytiotrophoblast and later the chorion of the embryo and the endometrium of the uterus; it may become much more intimate, involving the breakdown of maternal tissues so that the mother's blood bathes embryonic tissues directly. Even the outer layers of the embryo may break down to some extent, allowing the mother's blood to come in contact with deeper embryonic tissue. Nevertheless, the mother does not seem to produce an immune response to what is in effect a tissue graft, except in certain pathological conditions like rhesus incompatibility (see below).

It is now believed that there is no simple answer as to why maternal rejection does not occur. The uterus is not a completely **"privileged site"** (i.e. a site which fails to react immunologically), for tissues other than trophoblast *can* be rejected from it and embryos are not rejected when they develop **ectopically,** i.e. in the oviducts or even the peritoneum. Embryonic (even trophoblast) tissue, as well, *is* rejected if grafted onto the skin; but again this rejection is not as rapid by a pregnant animal as by a non-pregnant one, and the rejection is more rapid if adult rather than embryonic tissue is used to form the graft. The reason for this could be that the embryonic trophoblast is coated with a thin layer of protein which may isolate it; also, very early embryonic tissue does not express adult antigens. Furthermore, hormones, especially corticosteroids (which rise in pregnancy) are immunosuppressive and reduce the maternal response to any foreign graft. There are also odd bits of evidence that the mother *is* responding immunologically to the embryos: parts of the immune system of the mouse (notably the **metrial glands,** which are lymph glands next to the uterus) show a much greater response to genetically very different embryos than they do to embryos sharing most of the histocompatibility genes of the mother. It seems, therefore, that the mother *is* at least partly reactive and that the embryo *is* to a greater or lesser extent antigenic to her, but there is a delicate balance struck within the antigen–antibody system which makes the mother *somewhat* insensitive, and the embryo *not very* antigenic so that the threshold for rejection is not reached.

Although mammals do not normally respond to their embryos by rejecting them as if they were foreign grafts, there are other immunological interactions which may damage a developing baby. Of these, the best known is **rhesus incompatibility** in the human. Here the mother herself lacks an antigen (she is said to be *Rhesus negative*) which her baby may produce on its blood cells (it can be *Rhesus positive*, if the father is Rh+ve). The mother may become exposed to the baby's Rh antigen during or just prior to birth as placental tissues break down. *That* baby is not damaged, but the blood system of subsequent Rh+ babies may be attacked by mother's antibodies which normally diffuse into the baby's blood system without causing harm. The baby's blood cells may be completely destroyed, giving rise to a "blue baby". This was very often fatal to the baby, but mothers at risk are now given anti-rhesus antibody at the end of their first pregnancy, so that any of the embryo's rhesus antigen which enters her bloodstream is destroyed without causing her to make more of her own rhesus antibody.

DEVELOPMENT OF ORGAN SYSTEMS

Now that we have described the attainment of the phyletic stage by a variety of vertebrates, we are in a position to describe the development of the organ systems. Similarities among the different vertebrates in the early development of the organ systems are very close—only later do the characteristic differences of the adults appear. Therefore the descriptions will be divided into systems rather than animal types, and the various modifications of the basic plan characteristic of the different classes will be referred to briefly in each section.

The blood vascular system will be dealt with first, because its initiation is closely related to the basic anatomy of the embryo and because its development can easily be taken far enough that the student should make links with his knowledge of adult anatomy. The other systems to be described mostly depend upon blood vascular elements for their growth and organization and it is convenient to use blood vessels as geographical indicators in description.

Unfortunately, for the purpose of teaching **organogenesis** the various systems all develop at more or less the same time and correlate with one another to a considerable extent. Nevertheless, they must be described in sequence; a certain amount of cross-reference to other sections is therefore inevitable, although it has been kept to a minimum.

THE HEART AND VASCULAR SYSTEM

In the mesoderm of all vertebrates there are blood vessels. In almost all forms the arrangement of these vessels is related very closely to the **transverse septum.**

It will be recalled that the mesoderm of all vertebrates splits very soon after its movement over the edge of the blastopore, to become splanchnopleure (against the gut) and somatopleure (against the skin). Often, particularly anteriorly, the separation of the two layers is not easy to see; there are mesodermal cells against the skin and others against the gut, but there is a proportion in the middle which cannot be assigned to either layer. However, after the production of about a third of the mesoderm, splanchnopleure and somatopleure separate more completely; after this, coelom becomes more obvious between these layers posteriorly. The **cavity** in the anterior part of the coelom is much less obvious. After the appearance of the AIP the coelom is almost divided into an anterior region and a clear posterior area. In the chick, when the AIP (Fig. 39) folds the anterior part of the coelom (called the **pericardium**) against the posterior part (called the **peritoneum**), a kink in the ventral splanchnopleure results, forming the basis of the transverse septum (p. 95). Imagine a long "sausage" balloon being bent sharply but not very far; a crescent-shaped fold will develop on the inner side of the bend, partially separating the anterior and posterior parts of the balloon. Imagine the coelom is like the balloon, with the foregut running over and resting on the internal transverse fold developed as a result of bending (a bit like someone's head resting on the block of the guillotine!). When the coelom expands so that the gut is slung from dorsal and ventral mesenteries (see Figs. 40 and 47) the posterior part of the coelom around the gut (**peritoneal cavity**) is

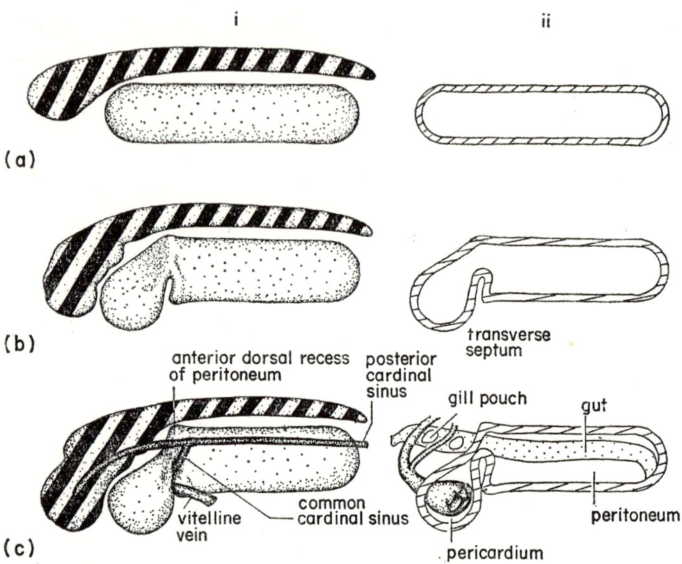

FIG. 39. *Development of the transverse septum. A very diagrammatic attempt to explain the formation of the transverse septum as a "kink" in the cylindrical coelomic cavity of vertebrates.* (*i*) *shows the nervous system and the coelomic cavity in the round.* (*ii*) *shows an approximate sagittal section of the coelom.* (*a*), (*b*) *and* (*c*) *show successive stages. In* (*c*) *blood vessels and the heart are shown; vitelline vein and common cardinal vein enter the transverse septum from each side* ((*c*)*i*) *and their "stumps" are shown on the heart in* (*c*)*ii. In* (*c*) *ii also the diagrammatic gut tube with two gill pouches is shown for orientation.*

narrow anteriorly (Fig. 39), where splanchnopleure and somatopleure are linked by the folded transverse septum. This is also, of course, the posterior part of a coelomic cavity, the future pericardium, surrounding the foregut and the heart below it. Its position is indicated in adult dogfish by the rear wall of the pericardium and in the mammals by the ventral part of the diaphragm.

The frog has no definable AIP (anterior intestinal portal) because its gut is an enclosed tube, with a floor, from its formation as the enteron. Splanchnopleure and somatopleure in the restricted part of the coelom now form the primitive **heart** and **pericardium,** by folding involving the ventral mesentery below the foregut. This is illustrated in Figs. 14, 33 and

41. In most vertebrates the heart is formed from a pair of rudiments, one each side of the midventral line under the foregut. DeHaan has shown that in the chick embryo these rudiments are first recognizable as groups of cells which crawl over the hypoblast from far laterally (⊕ Fig. 29). They seem to be oriented in their paths by differential "stickiness" of the upper surface of the hypoblast, so that the individual cells finally come to lie as a group on either side of the forming AIP. They already have a predilection for becoming heart cells (with an intrinsic rhythmic contraction) from the start of this long journey; this is shown by their behaviour in explants taken for tissue culture, where they form little beating aggregates. If their progress toward their destination is prevented by an obstacle, such as a gap in the hypoblast or an object in their path, they aggregate at the obstruction and make a more or less abnormal heart in the wrong place. It seems very probable that most telolecithal vertebrates form heart in this way; the situation in the mesolecithal forms is less clear; mammals probably resemble birds.

Meanwhile, areas of splanchnic mesoderm have been breaking down to give blood spaces. The cells lose cell contacts with their neighbours and soon begin to synthesize haemoglobin and become blood cells lying in tissue fluids. The areas run together forming **blood islands** and **sinuses** which wander over the forming gut and yolk sac and coalesce at the AIP into two channels, the **vitelline veins.** These veins run into the ventral mesentery under the foregut and so into the back of the heart (see Fig. 41). The blood then accumulates in the layers of the splanchnopleure and later the somatopleure, which bulge into the coelom between the pericardium and peritoneum along the fold of the crescent-shaped dorsal edge of the transverse septum, cutting the anterior and posterior parts of the cavity off from one another except for two ventral pockets and two dorsal passages (Figs. 39 and 40). The blood-filled bulges thus make a thick dorsal edge to the partition across the coelom. This is the *definitive* **transverse septum** and, in forms with an AIP (e.g. birds and mammals) it lies immediately behind the heart and immediately in front of the liver, which is forming anterior to the AIP from a diverticulum of the gut (see later, p. 103). The blood cavity becomes continuous in the transverse septum (Fig. 39) and the blood runs around the gut, joins the vitelline veins

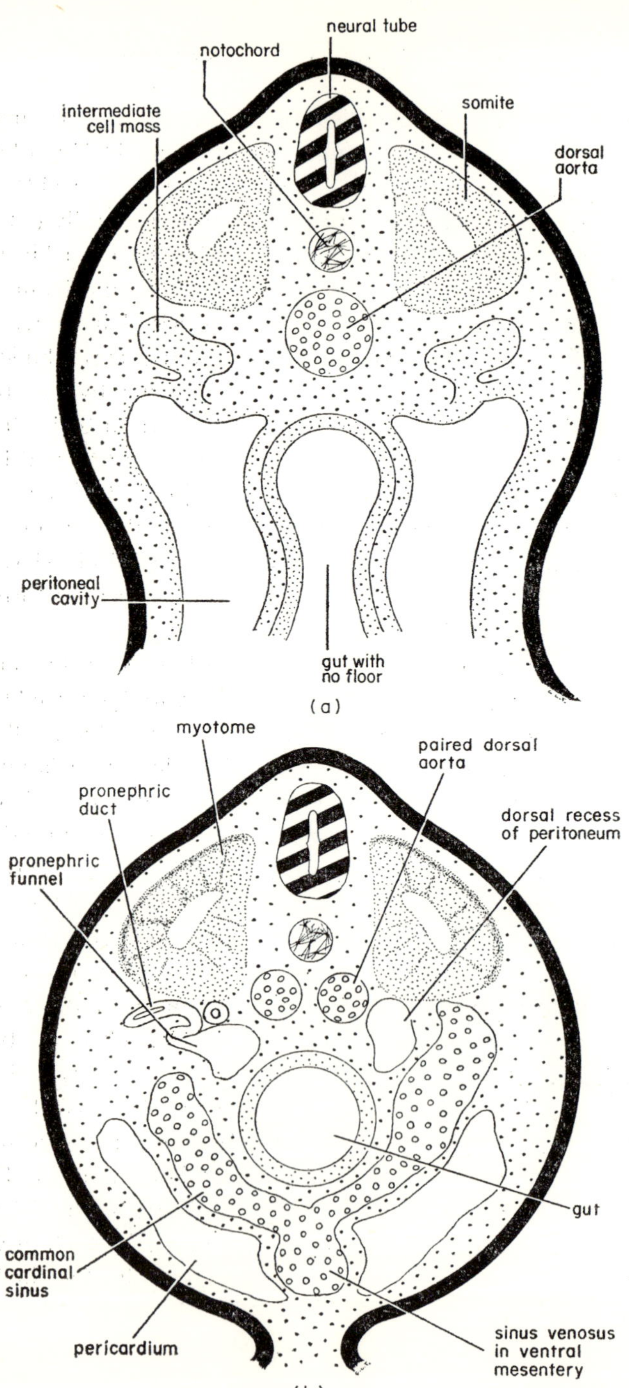
neural tube
notochord
intermediate cell mass
somite
dorsal aorta
peritoneal cavity
gut with no floor
(a)
myotome
paired dorsal aorta
pronephric duct
dorsal recess of peritoneum
pronephric funnel
gut
common cardinal sinus
pericardium
sinus venosus in ventral mesentery
(b)

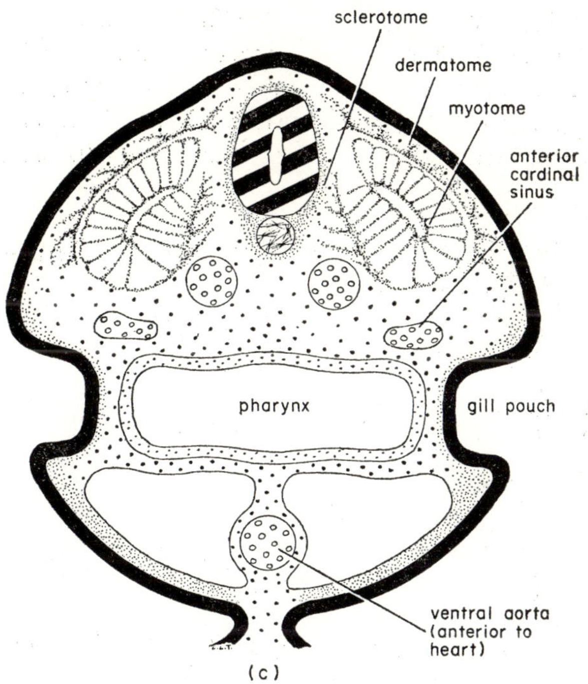

FIG. 40. *Transverse sections through a developing vertebrate. (a) in the abdominal region, well behind the AIP. The embryo (like a chick, selachian or mammal) has no floor to its gut and the large mass of dorsal mesoderm is undeveloped. (b) At the level of the transverse septum, more differentiation has occurred: pronephric ducts and funnels connect into the anterior dorsal recess of the peritoneum (see Fig. 39 (c)i). (c) In the pharyngeal region, gill pouches are shown and the somite has produced sclerotome and dermatome, but not, of course, nephrotome; those cells which would have become nephrotome aggregate with pharyngeal endoderm to contribute to thymus*

and enters the heart from the posterior end. The heart may have commenced to beat before any blood has poured into it, or it may wait until after the vitelline veins have entered it. The foregut has already commenced to form gill pouches (see below), which effectively block the passage of blood through them, so that blood which is forced from the front of the heart when it first beats can only penetrate upward *between* these pouches, forming the primitive **branchial vessels** (see Figs. 34 and 41). Above the gill pouches, blood runs forward into the "mesodermal porridge" of the head (which is mostly derived from neural crest, not true mesoderm) and backward as one or two vessels above the dorsal mesentery of the gut and below the notochord. The **dorsal aorta** is thus established, perhaps double at least at the anterior end. It runs posteriorly below the notochord but

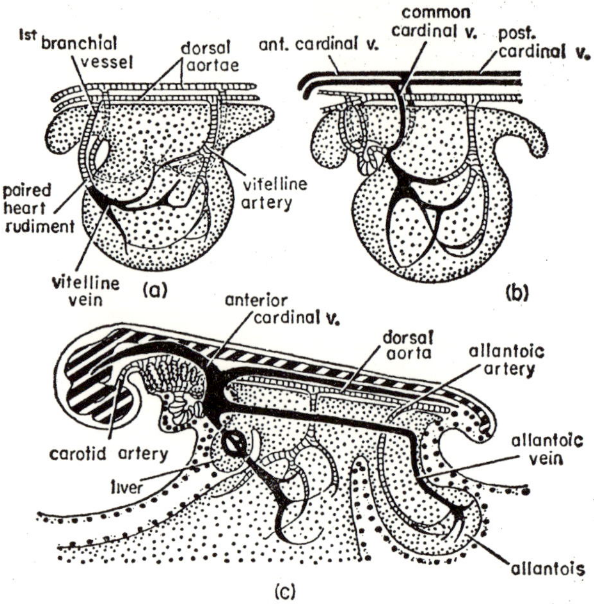

FIG. 41. *Development of blood vascular system in vertebrates. (a) The heart is a paired rudiment; dorsal aorta may be single or double. (b) The cardinal sinuses have appeared. (c) A full complement of branchial vessels and a liver circulation have appeared.*

above the dorsal mesentery until it comes to the blastopore or the neurenteric canal, where it *must* divide. The two branches are the **vitelline arteries,** which run around and into the splanchnopleure over the yolk sac and the gut, where they communicate with the blood islands again (Figs. 41 (a) and (b)).

The blood then runs forward in two main streams, one ventrally on either side of the gut where splanchnopleure originally formed blood islands. These follow the original vitelline veins, which enter the ventral mesentery (through the transverse septum) and so the posterior part of the heart. Meanwhile, more blood is being formed rapidly and this, combined with the increasingly effective heart beat, results in an increase in pressure throughout the developing vascular system. This pressure increase is very small, but it is sufficient to cause blood cells to permeate areas of loose mesoderm everywhere in the embryo, but particularly those on each side of the notochord (Fig. 40). In this way blood filled sinuses appear along each side of the embryo, lateral to the notochord. These are the **cardinal sinuses** which connect across the dorsal edge of the transverse septum and into the **sinus venosus** (vitelline veins fused anteriorly) above and behind the heart as described above (p. 95). The portions of the cardinal sinuses anterior to the transverse septum now have blood draining from the anterior, and are the **anterior cardinal sinuses.** The path into the sinus venosus, on each side (Figs. 39, 40 and 41), is the **common cardinal sinus (Ductus Cuvieri)** which marks the original dorsal edge of the transverse septum fold, linking somatopleure and splanchopleure. The posterior portions have blood draining the posterior parts of the embryo, and are the **posterior cardinal sinuses** (Fig. 41). Especially in the more posterior regions, the blood flow from the lengthening dorsal aorta to the posterior cardinal sinuses becomes restricted to the spaces between the somites, as the **vertebral arteries** and **veins.** This can be very clearly seen in the tail of embryo teleosts, as in Fig. 42. It will be recalled that the tail develops from the most posterior dorsal region of the embryo. Its first blood vessels may be a loop of the dorsal aorta, as in the teleosts (Fig. 42). Usually the two arms of the loop cross-connect by vertebral arteries and veins and the pair of these at the base of the tail may become the main bypass for blood to the yolk-sac. The ventral part of the loop then becomes progressively less important and may not remain patent; instead, blood from the originally dorsal loop finds its way into prolongations of the posterior cardinals, which may merge to a single vessel, the **caudal vein,** in the tail. In the tetrapods the aorta is usually drawn out as a single vessel, instead of a loop, and drains into tissue spaces which join up to the posterior ends of the posterior cardinals. These coalesce to form a main vessel, but often there are several drainages from the tail; for example, in frogs there is a

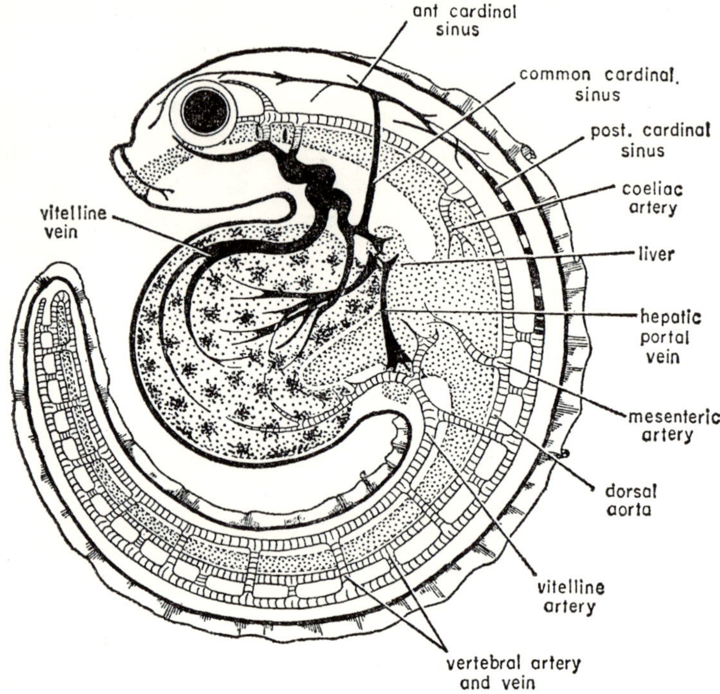

FIG. 42. *A teleost fish just prior to hatching. This diagram shows the relationship of the internal organs to the blood vascular system. (The so-called "yolk sac" is enclosed by periblast, not by endoderm.)*

mid-ventral one which proceeds forwards as the **anterior abdominal vein** and receives some of the blood from the hind limbs as they develop.

One further point must be considered here, although the foundation has not yet been laid. This is the subsequent history of the vitelline blood system as a result of the elaboration of the gut. The vitelline veins run along the anterior part of the gut (or over the anterior part of the yolk sac in embryos derived from telolecithal eggs). It is into these veins, below the foregut but behind the heart and transverse septum, that the **liver rudiment,** a ventral diverticulum from the gut, pushes. This outpushing breaks up the veins into smaller and smaller channels, with "fingers" of liver running between them. The posterior part of the vitelline veins, running over the gut, becomes the **hepatic portal vein** supplying the liver and the anterior part, from liver to sinus venosus, becomes the hepatic vein (Figs. 34 and 41).

Arteries run down the dorsal mesentery from the aorta to supply the gut and some of these survive in the adult as the anterior **mesenteric artery** (which was the old vitelline in tetrapods and now supplies the oesophagus, with a branch to the stomach), the **lienogastric artery** (to stomach and spleen), the **coeliac artery** (to the major part of the intestine), the **posterior mesenteric** and **rectal arteries**. The later fate of some other parts of the blood system will be described when we consider the later development of the various organ systems.

THE ALIMENTARY SYSTEM

It will have been realized that the initial formation of the gut tube must differ between homo-, meso- and telolecithal eggs and embryos. In the frog, as in other mesolecithal forms, the gut is a complete tube right from its inception. However, in telolecithal forms, like the chick, the gut forms as a pair of pockets: one is anterior to the AIP, the other posterior to the PIP; between the AIP and PIP the gut has no ventral floor. During development, these two endodermal pockets lengthen; that anterior to the AIP forms the **foregut,** that behind the PIP the **hind gut.** Between the AIP and PIP, where the gut still has no ventral floor, the **yolk-sac stalk** forms within the **umbilicus** and this connects the lumen of the gut with the yolk in the yolk-sac (see Fig. 33).

The anterior end of the gut tube approaches the ectoderm of the anterior end ventrally, and this ectoderm thickens and forms a pocket, the **stomodeum** (Fig. 14). Eventually, this pocket breaks through into the gut and a true mouth is thus acquired. Meanwhile, just posterior to this in the pharynx, lateral bulges of the anterior gut tube have also come into relationship with ectodermal inpushings, forming the **gill pouches** (Fig. 43 and Plate XI b, d). We have seen that the gill blood vessels run up in

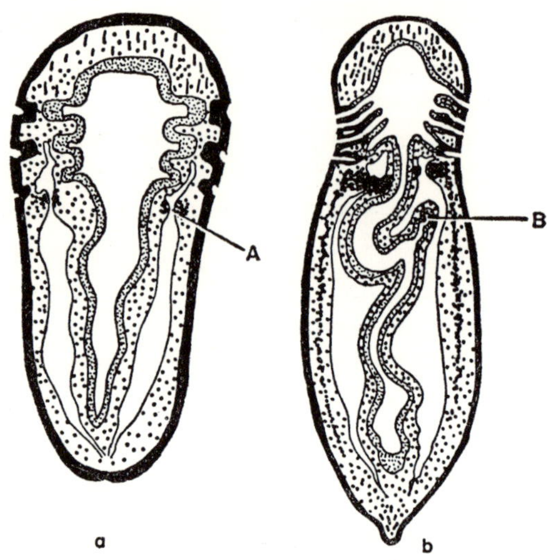

FIG. 43. *Horizontal sections of young tadpoles. (a) Before the gills have broken through. Note the narrow part of the coelom behind the last gill pouch, engorged with blood (A). (b) The gill pouches have broken through, as have the common cardinal sinuses, the stomach is bent and the bile duct (B) is moving forward.*

the mesoderm between these pouches. The gill pouches only result in functional gills in the fishes and the larvae of amphibia; in all other vertebrates, they become modified in other directions and the first or **hyoid** pouch contributes considerably to the ear. The dorsal part of each of these gill pouches is nipped off into the mesoderm as a small mass of tissue, lymphoid in appearance. These, which contribute to the thymus or parathyroid glands, are the **supra-branchial bodies.**

Still further posteriorly, just above and behind the transverse septum, the primordium of the lungs or swim bladder appears, ventrally or dorsally respectively. This pushes the splanchnopleure before it so that the lungs, for example, are sheathed in connective tissue continuous with that of the trachea and pharynx. The cavities of the lungs and of some swim bladders remain connected to the pharyngeal cavity, but in some fishes the swim bladder seals off and moves posteriorly (this occurs in all the teleost fishes mentioned here).

In the mammal the splanchnopleure carried on the posterior end of the lungs contributes to two dorsal segments of the diaphragm; connective tissue and coelomic epithelium then grow in and cut off the lung from the forming diaphragm so that it lies relatively free again.

Further behind the transverse septum, the development of gut becomes very complicated. Let us consider this under three headings:

(a) *Development of the Liver*

The liver arises, usually from two outpushings of the gut tube, about midway between the AIP and the transverse septum (see Figs. 41, 42 and 43). One will give the **gall bladder,** the other the liver itself; the **bile duct** arises from either or both. The liver primordium pushes out ventrally into the splanchnopleure of the ventral mesentery (see Figs. 9 and 42 and Plate XIe) and then expands in this mesentery and forward into the thickness of the transverse septum, breaking the vitelline veins into capillaries and so establishing the hepatic portal system. Later, the liver separates again from the transversc septum by ingrowth of peritoneal epithelium, remaining attached only by the **median hepatic ligament,** which contains the hepatic vein. The ventral mesentery persists above the liver as the **hepatoenteric ligament** between the gut and the liver. It also persists below the liver, anchoring it to the ventral body wall as the **falciform ligament,** extending to the umbilicus. Note that the bile duct, hepatic portal vein and the hepatic artery (a branch usually of the coeliac artery) run in the hepatoenteric ligament (see Fig. 44).

(b) *Development of the Pancreas*

The **pancreas** arises from several outpushings near the origin of the bile duct and it spreads anteriorly over the forming stomach.

(c) *Development of the Stomach and its Supporting Mesenteries*

The anterior end of the stomach is anchored by the oesophagus in the transverse septum; the posterior end is also anchored via the bile duct and hepatoenteric ligament. Therefore, as the stomach extends in length, it has no choice but to curve, always into the left half of the anterior peritoneal cavity (Fig. 43). It is suspended by dorsal mesentery, like the rest of the gut. This mesentery follows the curve of the stomach and is so thrown into a fold. This fold seals as the two parts of the mesentery touch and forms the **mesogastrium** or **omentum.** In this run the coeliac artery and later the **inferior vena cava.** It forms the dorsal edge of the **foramen of Winslow** (Fig. 44).

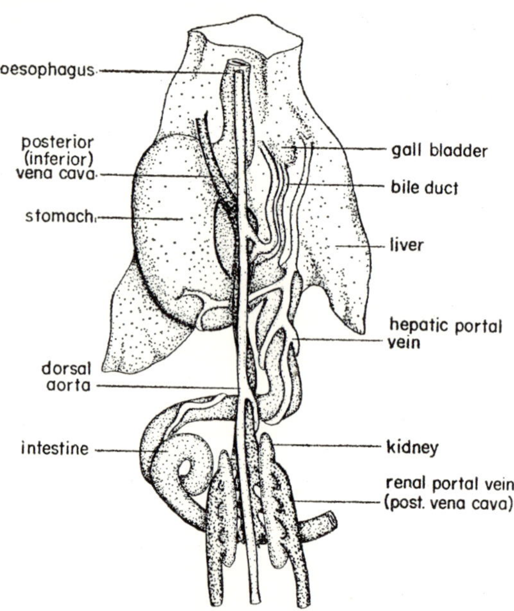

FIG. 44. *A dorsal view of the contents of the peritoneal cavity of a young vertebrate. Note: in previous editions we have shown the mesenteries connecting all these organs, but students have told us that this made the diagram too complicated to understand; accordingly they have been omitted this time.*

In mammals the yolk-sac stalk is closed off at birth; in other vertebrates (e.g. dogfish) the entire yolk sac is absorbed and its endodermal lining completes the floor of the gut.

Posterior to the PIP the allantoic stalk runs forward to fall over the PIP into the umbilical cord. Its fate varies from group to group and it will be considered more fully under the urogenital system.

The hind end of the gut, primitively open at the blastopore (but not in birds, or mammals other than primates), loses this connection with the "outside world" as the neural folds arch over and close the blastopore; instead, it becomes confluent with the neural canal via the neurenteric canal. An ectodermal invagination approaches the hindgut, ventral to the old position of the blastopore; this will form the **proctodeum** (Figs. 14 and 42).

In the mammal, the anterior edge of the allantoic stalk (= the posterior lip of the yolk-sac stalk) *appears* to grow backwards as a shelf until it hits the **cloacal membrane** where hindgut endoderm and proctodeal ectoderm meet. No tissue, in fact, progresses posteriorly, but the whole procedure is accomplished by transforming the **O**-shaped gut into an **8**-shape progressively towards the posterior end (see Fig. 51). The shelf thus formed, the **urorectal septum,** divides the hindgut into ventral **urogenital sinus** and dorsal **rectum.** This will be considered further when discussing the urogenital system.

The characteristic histology of the various parts of the alimentary system appears as the parts become morphologically distinct. It is usually retarded relatively in those epithelia which contain much yolk, for example the midgut of the tadpole and the resorbed yolk sac of fishes. There is some evidence that the regional differences of epithelium are not primary but are responses to differences in the underlying splanchnic mesoderm. This will be considered further when we deal with cell differentiation (see pp. 136–143).

THE NERVOUS SYSTEM

The induction and early development of the primitive neural tube have already been described for the frog and are similar in most vertebrates except the teleost fishes. The changes which occur in the gross morphology of this primitive tube to become the definite central nervous system are also remarkably uniform among the vertebrates.

At the anterior end, above and anterior to the pharynx, the neural groove becomes much wider as the brain divisions and primordia of the eyes appear (Figs. 13, 30 and Plate Xc). Here the neural groove "falls over" the anterior end and so, when it closes to form forebrain, there is a flexure sometimes as much as 90°. This, the **cephalic** or **cranial flexure,** rotates the anterior coelom, the pericardium, to make the transverse **septum** (Fig. 39). It is subsequently corrected in most vertebrates, either by a growth differential in the head region (e.g. dogfish) or by a compensating flexure at the junction of head and neck (most mammals); we retain the cephalic flexure into adulthood. The division of the brain into **fore-, mid-** and **hindbrain** may be clearly seen in most embryos (Fig. 31 and Plates VIIIe, Xc). The floor of the forebrain is in intimate contact with the stomodeum; it is in this region that the **pituitary gland** is formed by association of components from both.

The cells within the wall of the spinal cord have a special way of developing the histology and circuitry of the adult nervous system (see Fig. 45). In the neural plate, which is only 1 or 2 cells thick to start with, the individual cells elongate so that the plate thickens greatly; each cell retains its connection to both the inner and outer aspects of the neural plate. Because all the cells are long and thin and the region containing the nucleus is the widest, nuclei appear to be scattered through the thickness of the plate, as if there are many layers of cells; but this is not the case. When the neural tube closes most of these cells undergo mitosis: the nucleus of each cell drops towards the inner surface of the tube (the luminal surface) and undergoes a transverse division. One of the resulting cells retains its "footing" at this surface, but the other daughter "escapes" towards the outside of the spinal cord. As it proceeds away from the luminal surface it differentiates, until by the time it is near the outer surface, some days later, it has dendrites and an axon. If the cell is in a part of the spinal cord next to a somite its axon grows out to innervate this; if it is in the dorsal part of the cord, where fibres from the dorsal root enter, they connect to it. However, if its axon fails to make any connections and it does not become involved in the internal circuitry of the spinal cord itself, then it dies. The **brachial** (arm) **plexus** and the **sciatic** (leg) **plexus** are associated with bulges in the spinal cord, not because

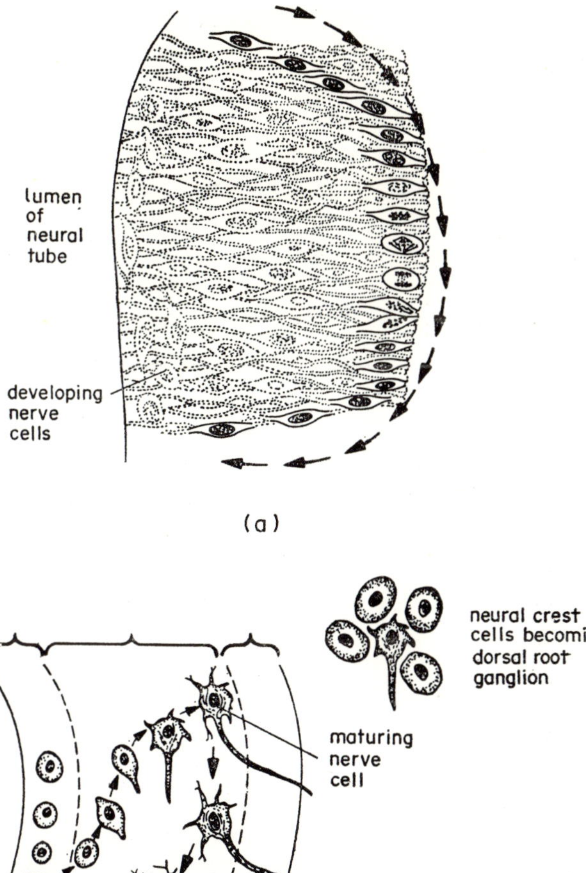

FIG. 45. *Development of neurones in the wall of the spinal cord.* (*For explanation see text opposite*)

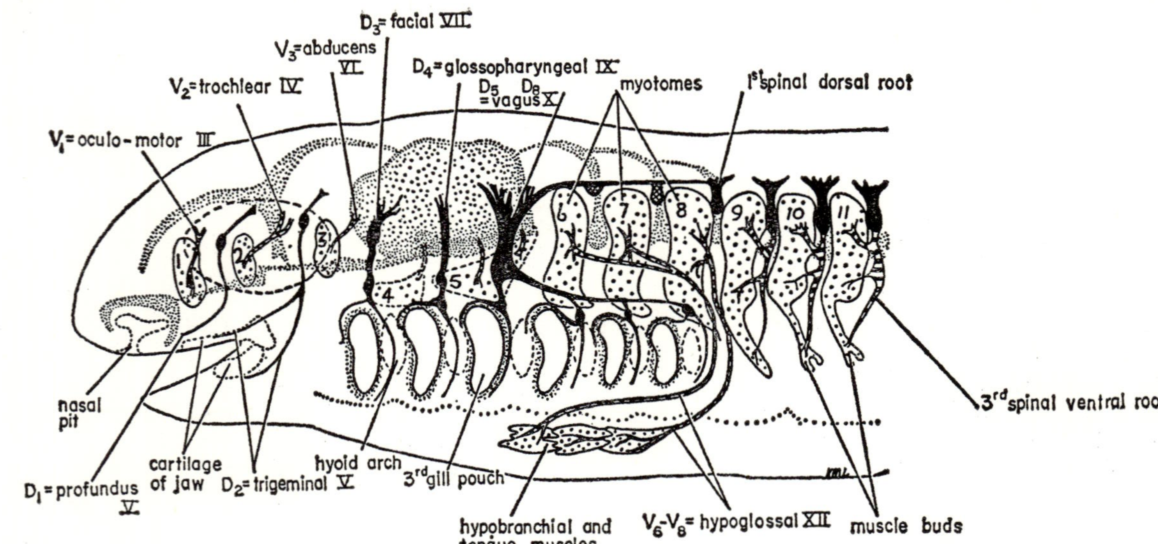

FIG. 46. *Developing head of the dogfish embryo, modified after Goodrich, to show the basic segmentation.* V_1—V_8= *ventral roots (motor) of cranial nerves.* D_1—D_8 = *dorsal roots (sensory) of cranial nerves. The equivalent traditional labelling is shown in Roman figures. Somites 4 and 5, and their ventral roots, "disappear" as the ear develops.*

more cells have been produced there but because fewer cells have died. If an early limb bud is transplanted to a region between fore and hind limbs of another embryo it "saves" the spinal-cord cells of this region and a larger diameter plexus is found here as well.

Along the whole length of the vertebrate head and trunk and usually tail, the mesoderm on either side of the spinal cord and brain becomes divided up into a series of somites (see above, p. 66, and Fig. 47). Each somite induces the production of a **ventral root nerve** from the nerve cord where it contacts it. This results in the appearance of **segmental** ventral roots, corresponding in number and position with the somites (Fig. 46). In the head region, especially of the tadpole of the common frog, the

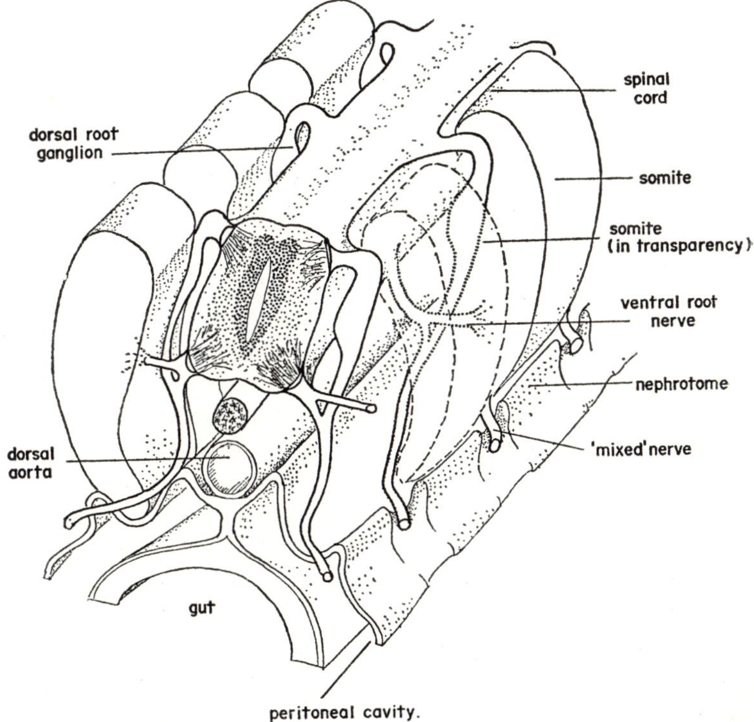

FIG. 47. *Part of the trunk of a vertebrate embryo shown dissected. Note dorsal roots are intersegmental, ventral roots segmental*

somites are not clearly demarcated, but the position of the ventral roots indicates the segmentation. Many of the myotomes (muscle blocks developed in the somites in response to the nerves from the spinal cord) are destined to move very considerably during development; in these cases their origin may be clearly established by tracing the motor nerve (ventral root nerve) back to the central nervous system (e.g. those of the mammalian diaphragm come from cervical roots).

With the development of the dorsal root ganglia, the situation becomes rather more complicated and we deal with this after consideration of the early development of the muscular system and, particularly, the neural crest (p. 112).

THE MUSCULAR SYSTEM

The muscular system of vertebrates arises partly from the segmental somitic mesoderm, partly from the more ventral unsegmented **(lateral plate)** mesoderm and partly from ectomesenchyme derived from neural crest cells.

Along each side of the spinal cord, and usually of the brain, is a row of mesodermal somites. As these develop each becomes divided, probably under the inductive influence of neural tissue medially, skin laterally, and perhaps gut ventrally, into four regions (Fig. 40). These are **myotome** (from which most of the musculature will arise), **sclerotome** (lying close to the notochord and destined to form the vertebral column and the proximal ends of the ribs), **dermatome** (lying under the skin, to whose dermis it contributes) and an **intermediate cell mass,** soon becoming **nephrotome** (kidney primordium) and **gonotome** (gonad primordium). The actual **germ cells** migrate up the dorsal mesentery into this tissue from where they have been sitting in the gut endoderm. This has been shown in mammals, in birds and in an amphibian, mostly by irradiation of various primordia and localization of the resulting defects; irradiation of early endoderm leads to germ cell damage.

All of the segmental muscles of vertebrates arise in the embryo as paired concentrations of somatopleure on either side of the nerve cord and notochord, the myotomes (Fig. 47). The situation in the head seems complicated (Fig. 46), but can be understood by comparing nerves, somites and gill structures. Myotomes of the three most anterior segments provide the extrinsic eye musculature, myotomes four and five degenerate more or less completely; six, seven and eight contribute to the gill musculature and may perhaps grow ventrally and contribute to the muscles of the lower jaw and even the tongue (Fig. 46). The diaphragm of mammals contains muscles derived from cervical segments. The limb and fin muscles are prolongations from a series of myotomes relating to the original **limb bud:** tongues of denser mesoderm from the somites in the limb region come to form a limb bud. These tongues become "squashed" together causing the bud to change in shape from a crescent to that of a very dumpy limb, as the base contracts around them. Perhaps as a result of this, many of these mesodermal tongues become muscle fibres which soon attain junctions with the ventral root nerve of their somite. They begin to contract spasmodically, bending the limb bud; where it bends, a joint develops with its capsule; proximal and distal to this, where the loose mesoderm between the muscles is squeezed and compressed rather than bent, cartilage is desposited which lays the foundation for the limb bones.

The more ventral somatopleure (lateral plate mesoderm) remains unsegmented and the muscles which arise in its substance are frequently longitudinal. In the head the muscles of the jaws, the tongue and the gill arches are all mainly derived from lateral plate mesoderm. In the trunk, most of the muscles of the ventral body wall derive from myotomes, but some of the longitudinal ones condense as sheets of muscle cells in the lateral plate.

The contributions to the musculature from the neural crest will be considered under that heading.

THE NEURAL CREST

This is one of the most important organ-forming systems in the embryo. It derives from that epiblast or outer layer just beyond the strong inductive influence of the prechordal plate and notochord and hence, when the neural folds form, it is found on their lateral and dorsal aspects. As the tube closes, the neural crest cells drop into the mesoderm as **ectomesenchyme,** only distinguishable from local somatopleure with great difficulty. By this time the somites have become divided into myotome, dermatome, etc.; a loose mesh of connective tissue forms between separate successive myotomes. These intersegmental partitions, **myocommata** (singular: **myocomma**), provide a major pathway for the neural crest cells to invade the deeper tissues of the body.

Other important pathways along which neural crest cells migrate, in addition to the myocommatal route are the dermal route (to skin and other superficial parts), the neural route (around the nerve cord and along the developing nerves) and the visceral route (around the nerve cord and notochord into the dorsal mesentery and thence onto the gut and its derivatives, the liver).

Above each myocomma the neural crest cells aggregate as a **dorsal root ganglion** which sends nerve processes into the spinal cord. Note that the dorsal root ganglia are *intersegmental* and, in fact, they each lie *behind* the segment which they innervate (Fig. 47). After establishing connection with the contiguous spinal cord by the **preganglionic fibres,** the cells of the dorsal root ganglion send axons to establish sense organs, not only in their own segment but frequently (as in the lateral line system or taste bud system of many fishes) over the general body surface. The main branch from the ganglion passes down the posterior face of its myotome. In the trunk region, this main branch soon joins the ventral root nerve of the myotome to form a typical mixed spinal nerve, with a sensory component from the dorsal root and a motor component from the ventral root. Frequently, however, the relationship (dorsal root = sensory; ventral root = motor) is complicated by other factors, especially in the head region. In many cranial nerves the dorsal component may contain motor fibres and the ventral component may contain sensory fibres; e.g. the dorsal roots of the vagus nerve have motor fibres which influence heart rate. The autonomic nervous system also originates entirely from the neural crest. Some of the cells which take the visceral route aggregate below the notochord as the sympathetic ganglia and processes from these pass down the mesentery and innervate the gut at the same time as other neural crest cells are invading gut tissues. (It is just possible that the various kinds of endocrine activity of the gut region are mediated by the neural crest cells rather than the endoderm cells themselves or their associated mesoderm.) At this time, neural crest cells invade all the tissue of the body as well as the gut; for example, they arrive at the wing bud of the chick embryo at about 72 hours, compared with 60 hours for the mid-gut. In the developing wing they contribute, among other things, to the pigment cells (see p. 117).

The **medulla** of the **adrenal gland** of mammals and the **suprarenal glands** of other vertebrates is, like the sympathetic ganglia, neural crest in origin; so are the **Schwann cells** ensheathing nerve fibres. So is all of the special pigment cell system which results in the melanin pigmentation of hairs, feathers, skin and often of internal organs too (see p. 117). Another important contribution of the neural crest is to the **branchial arches.** These are cartilaginous supports running behind each gill pouch. The hyoid gill bar makes the tiny ear bones of tetrapods (see p. 123). If the neural crest of one side is removed, then not only do no dorsal root ganglia appear in the affected area, but the area does not receive its complement of pigment cells and no gill bars are produced in it.

The neural crest cells are in general a migratory group and, while much of their subsequent history has been discovered by extirpation experiments, it is certain that many of their final sites remain unsuspected. It is known that the pigment cell precursors (melanoblasts) penetrate into all tissues of the vertebrate body and that their tissue environment determines whether or not they shall transform into pigmented melanocytes. The extent of contributions by neural crest cells to, for example, the intermediate lobe of the pituitary and to the spinal membranes is doubtful. The undoubted contribution to the mesenchyme of the whole head region tends to disturb the results of extirpations in this area, so that these tissues or organs could not develop normally in any event. It is probable that in the head region of most vertebrates there is a major contribution by the neural crest to the muscles and connective tissue of the head. The myotomes from the first three somites contribute only to the extrinsic eye muscles; four and five may contribute muscle in the roof of the mouth of snakes and lizards and other forms with very "articulated" skulls like the teleosts. The rest of the connective tissue dorsal/anterior to the mouth is mostly neural crest derived, but it forms dermis, the dermal papillae and sheaths of hairs, feathers, vibrissae, and scales just like the "proper" mesoderm of the trunk. It also forms the muscles of these structures (e.g. the arrector pili of hair follicles) and probably the circular sheet of muscle which lies in the iris and contracts the pupil of the eye. Probably, too, it makes the dermal pulp of the teeth, at least in the upper jaw.

THE SKIN

The skin of vertebrates consists in the adult of two major layers, the **dermis** and the **epidermis.** The gross morphology of their development is relatively simple, but their interactions are very complex and only a simplified account can be given here.

The epidermis usually persists from the outer layer of blastular cells as a 1- or 2-cell-thick layer covering the entire embryo except, of course, the blastopore, or primitive streak if this persists (as on the cloacal membrane of mammals, p. 105). Usually coincident with the beginnings of organ development the cells divide to form an outer layer, the **periderm.** This plays no further part in the construction of the morphology of the epidermis and is cast off during development. Much later, when most of the internal architecture has been completed (16-day mouse embryo, 5–6-day chick), another wave of cell divisions in the inner layer results in a succession of cell layers above the original, which remains as the **basal layer;** the typical squamous epithelium is thus established. In the fishes and the aquatic stages of amphibians, it remains in this state, but in the tetrapods it dries out in its outer layers and these become specialized to form **keratin;** these layers are constantly renewed from the basal layer during life, so that the skin exhibits a kind of permanent embryology. As will be seen, this applies to its maintenance of regional character, as well as to its maintenance of its cell population.

The dermis of the trunk derives most of its cells from the dermatome part of the somites, especially more dorsally; ventrally, the **lateral plate** somatopleure forms dermis. The dermis of the head region, on the other hand, may be formed partly or even completely from neural crest cells. However, there seems to be little difference in development. The dermis starts as a fairly dense layer of basophilic cells lying immediately under the epidermis. A **basement membrane** is soon secreted, perhaps cooperatively by both layers; this changes considerably in its staining reactions during development, probably indicating different degrees of interaction between the two layers across it. Just prior to the epidermis becoming many-layered, the dermis thickens, at least partly because of the secretion by its cells of fine fibres which lie between them. After the epidermis has attained virtually its adult structure, cells in the dermis lay down **collagen** fibrils, which aggregate in very specific patterns to give the characteristic pattern of that kind of skin (and the quality of the leather it may later produce). In the cornea of the eye these aggregates, the collagen fibres, align parallel to the surface in a very regular mesh which in part accounts for the transparency of this part of the dermis.

Apart from the irregularities caused by the growth of various epidermal appendages (e.g. hair, feathers, teeth) the junction between dermis and epidermis is usually wavy, especially in the mammals. The upward projections of the dermis into the epidermis have been called "dermal papillae", but this is a misnomer as they are almost always long ridges rather than papillae (they define the ridges of fingerprints, for example). The term "dermal papilla" is best reserved for the very highly specialized organ which projects into, and organizes, the hair or feather germ (p. 116).

The regional character of the epidermis (which is shown by difference in thickness of the various layers, in hardness, and in permeability over the surface of one animal) are *not* intrinsic to the epidermis itself. The specific character is constantly maintained by the action of the underlying dermis, as may be shown by transplantation experiments: epidermis always assumes the character of the area from which its dermis was taken, except in the very early embryo. Reciprocal interactions occur during the initiation and early development of the skin and its appendages. Most of the experimental work has been done with the feathers of chick embryos, but the results seem typical of the general story:

(1) The presence of the axial organs (e.g. notochord, spinal cord) inside the skin cause the initiation of changes, probably first in the epidermis, which result in the first, mid-dorsal, pteryla. (Other solid objects, even chips of aluminium, which stretch the skin from inside have a similar effect.)

(2) Slight aggregations of cells appear in regular positions in the dermis. (Young epidermis placed over this dermis forms feathers over these aggregations. Therefore, the dermis is now affecting the epidermis.)

(3) The epidermis above these aggregations forms slight bumps and thickens. (This epidermis, when placed on younger dermis, causes dermal aggregations and feathers in these positions. Epidermis is now affecting dermis.).

(4) The aggregates now condense into recognizable **dermal papillae,** the active mesodermal cores of the future feather follicles. (These produce intensive cell division in younger epidermis placed above them. Specific parts of the dermis, the dermal papillae, are now affecting epidermis. Further, the rest of the dermis controls the differentiation of the general surface epidermis.)

(5) The epidermal bump, having grown taller and become cylindrical, causes the proliferation of cells from the outer side of the dermal papilla to form the dermal **pulp,** which fills the inside of the cylinder as it grows.

The sinking of the base of the cylinder into a pocket in the skin, the **follicle,** is also doubtless controlled by similar interactions. The changes in the form of feathers as the bird grows up, and the production of a series of feathers from the same follicle, are all known to be controlled by a complex series of interactions between dermis, epidermis and dermal papilla during the whole life of the bird. The growth of hairs, nails, horns, scales, and the whole variety of epidermal glands is certainly similarly controlled. The actions of outside influences, for example hormones, on this system often appear contradictory, but this is only to be expected when such complex systems are controlled. For example, male sex hormones cause much bigger hairs to be produced from the beard region in men, yet the same hormones cause male baldness! To make matters worse, topical application of the same hormones may apparently reverse the changes of male baldness!

There is another cell population in the skin which we have not yet considered, but whose effects are at least as dramatic. This is the **pigment cell system.** Cells from the neural crest invade the epidermis, first dorsally and then progressively to the mid-ventral line. In the species where some of them have begun to make pigment in the early stages of their journey (a few fishes and amphibians) their progress may be traced easily. In most vertebrate species, however, pigment production does not start until the tissues in which the pigment cells lie have differentiated. In these cases, other methods of investigation must be used. If bird tissue containing pigment cells is implanted into the wing bud of a 72-hour chick embryo before its own neural crest cells have arrived, these cells migrate out, divide many times, and invade the forming feather follicles. When the chick hatches, its wing may be partially or entirely pigmented by the implanted cells. If the implant is made later, though, the patch of donor pigmentation is much smaller, and implants into wing buds of embryos older than 85–90 hours do not contribute pigment to host plumage at all. This is because the wave of the host's own neural crest cells has populated the wing bud by then. Tissue culture of wing bud of various ages, or grafting into the coelomic cavity of white breed embryos, confirms this: grafts from embryos up to 72–80 hours old form only white feathers, older grafts show good normal pigmentation.

It must be noted that many, perhaps most, of the pigment cell population from the neural crest are fated never to produce pigment; only those cells which chance to lie in a region of the animal which gives the "right cue" for pigment production will ever actually produce pigment. The others will remain in the tissues throughout the life of the organism, without exhibiting their potential. They can be shown to be present (although normally indistinguishable from other cells) by culturing the tissue in an appropriate tissue culture medium; some of them may then form pigment. The blanket term for those cells which possess, or whose descendants will possess, pigment-forming ability but have not yet made pigment is melanoblasts; cells which make pigment and store it (like those in the dermis of amphibians and reptiles) are called **melanophores;** those cells which make pigment and donate it to nearby (epidermal) cells are called **melanocytes.** This cell system is discussed again when we come to cell differentiation (p. 141).

The detailed development of hair, feather, nails and teeth is not dealt with here as it seems appropriate to a more detailed text.

SENSE ORGANS—PREAMBLE

The sense organs of vertebrates with which we can deal are the eye, the ear and the nasal organ; many other kinds of sense organs, for example the lateral line of fishes and larval Amphibia, the sense organs of chemical discrimination, and proprioceptive organs, are not clearly separable from surrounding tissues and their development cannot be considered here.

The eyes develop in association with part of the neural plate; the ears develop from a variety of structures in the posterior part of the head; and the olfactory epithelium is probably best considered as the very front edge of the neural plate itself. The development of these organs is of especial interest as each shows, in a different way, the manner in which contributions from very different sources combine to make functional units. Other organs, most notably the mammalian diaphragm (p. 103), derive from a variety of sources, but this does not seem to be reflected in their function as clearly as in the three sense organs now to be described.

THE EYE

The neural plate widens anteriorly (perhaps due to the strength of prechordal plate as organizer) and its lateral edges form prominent pockets on its side walls when it closes up to enclose the brain cavity (Fig. 13(d)). The original lower surface of the neural plate has been folded up to form the outer sides of the neural tube and is, therefore, now pressed against epidermis (or perhaps neural crest) from the underside. These pockets enlarge and form the **optic vesicles,** while the epidermis over them thickens and becomes the **lens placode.** As the optic vesicles enlarge they do not remain hemispherical; instead, their upper aspect becomes flattened and their undersides become grooved (Fig. 48). As they continue to grow, the ventral aspect "collapses" further into the upper wall of the cup to make a shape a little like a squashed tennis ball (Fig. 48). Each is now more or less cup-shaped, but with a double wall, and they are known as the optic cups. The rim of the cup now becomes complete by the walls of the groove growing together and this line of fusion moves inwards from the margin of the cup towards the brain. However, a gap persists near the base of the cup where the last bit of the ventral groove fails to seal up; this gap is called the **choroid fissure** and is very important as a way into and out of the eyeball for blood vessels and nerves.

The lens rudiment has meanwhile thickened and taken up its position in the mouth of the cup. There is clear evidence that in most vertebrates the lens is induced to form in the overlying epidermis by the action of the underlying optic vesicle: transplanted optic vesicles can induce lenses to form from almost any local epidermis, even on the flank. As the lens drops into the mouth of the optic cup, it becomes more spherical, and its internal cells rearrange to form an outer layer with a thickening on the back wall; these cells now produce or become **lens fibres,** which account for most of the substance of the adult lens and for its optical properties. The edges of the optic cup become the **iris** and develop a ring of muscles (Fig. 49). The inner wall of the cup becomes the **retina** and its innermost layer of cells produces axons which creep around the inner surface of the cup and out by the the choroid fissure. The cup is still joined at its base to

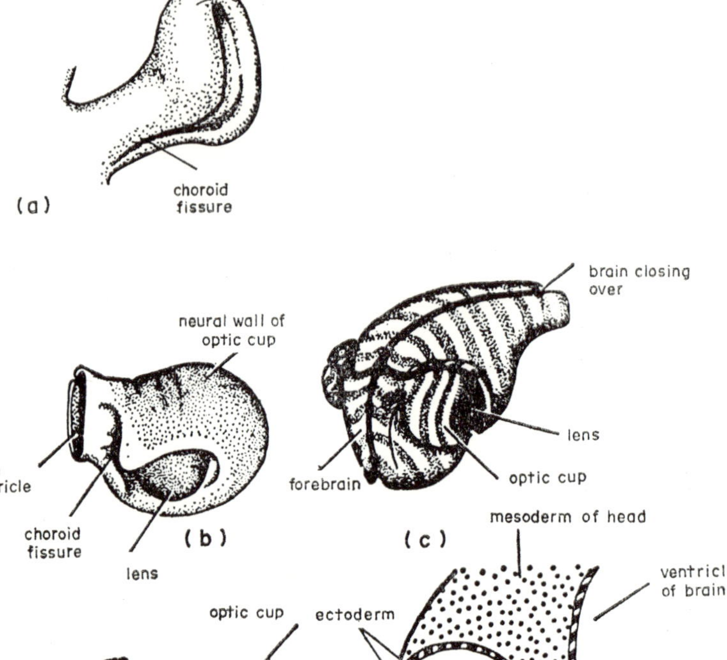

FIG. 48. *Development of the eye.* (*a*) *Diagram of the developing optic cup.* (*b*) *Drawing of a dissection of an early eye cup.* (*c*) *Diagram of the forebrain and two optic cups, with their lenses, at a stage comparable with* (*b*). *The arrow points to the choroid fissure.* (*d*) *Longitudinal half of a comparable stage with* (*b*). *The arrow is meant to show continuity in the extension of the brain cavity.* (*e*) *The appearance of this optic cup in a section of chick head at about 80 hours incubation; the inevitable choroid fissure indicated by the arrow.*

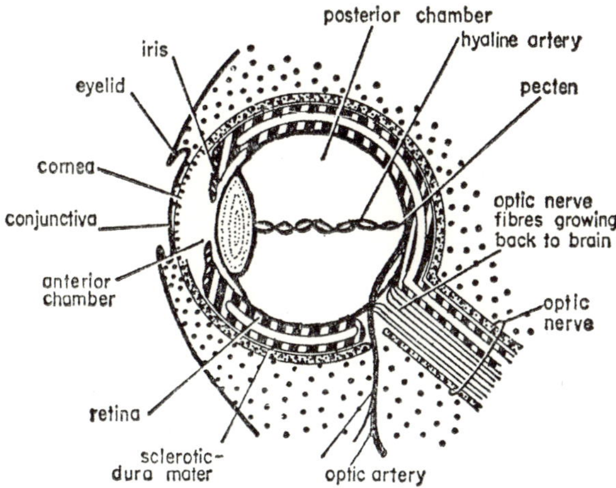

FIG. 49. *Section of the formed eye to show the embryological derivation of its parts.*

the mid-brain by the **optic stalk,** along the underside of which these axons find their way into the brain. These axons form a bundle, the **optic nerve,** which dwarfs the original optic stalk which itself persists as a thread along the dorsal side of the nerve. Half of these axons (in man) pass over to the opposite side of the brain and half remain to form connections on their own side; the place where the fibres from the optic nerves of opposite sides of the brain cross over each other below the midbrain is called the **optic chiasma.**

As organs grow in the cellular porridge which fills the head the loose mesenchymal cells are layered onto their expanding surfaces. Over the nervous system, including the developing eyes, a thin layer of cells forms a tenuous sheet. This becomes the **pia mater** over the spinal cord and brain, and a similar layer invests the early eye-cup, forming its boundary so that the nerve fibres from the inner surface of the future retina must slide around under this to the choroid fissure. This layer keeps them closely applied to the retinal surface and the orientation of the cells in the investing layer may guide the axons toward their point of exit from the eye-cup. As the lens drops in, a similar investment forms the lens capsule. The heart has now begun to pump blood and channels (sinusoids) run in the head mesenchyme, which now has a fibrous matrix secreted by its cells (reticulin fibres). As the organ grows it squeezes the loose connective tissue around it; as sinusoids develop they tend to be "squashed" on to the surfaces of the eye cup, so that these come to have a plexus (layer) of small blood sinuses. Around the brain and spinal cord, this is known as the **arachnoid coat,** and it is expanded on the dorsal side of the brain to form the **medullary** and **choroid plexus;** but it also forms over the inner and outer surfaces of the optic cups and remains as a plexus of **retinal blood vessels** into the adult. As the lens occludes the outer aspect of the cup, the plexus forms a ring around it in the plane of the iris, part of the **Zonule of Zinn,** and does not cover the surface of the lens or the inner aspect of the future cornea. It will be seen that this investment of the optic cup/lens system as a unit gives a blood supply to the inner surface of the retina (fed from the mesenchyme outside through the choroid fissure), but does not involve blood vessels in the areas used for refraction.

As the dermis of the skin begins to form collagen fibres, the rest of the mesenchyme of the head also becomes more fibrous, especially over the surfaces of organs. The organs are still growing, so another fibrous layer is laid down over them (imagine inflating a balloon in a loose fibreglass mesh—as it expands a fibrous sheath layers onto the surface). The **dura mater** of the brain and spinal cord is thus laid down and the **sclerotic coat** of the eyes. The choroid fissure is now almost filled by blood vessels and nerves, and the anterior chamber area (Fig. 48) outside the lens is almost free of cells, so the sclerotic is laid down evenly over the outer surface of the eyeballs, except for the area in front of the lens and behind the cornea. The **cornea** itself has a very strange kind of deposition of collagen fibres, all parallel to the epidermis in a mesh quite unlike that of dermis of the general body surface. It seems that this arrangement, which increases the clarity of the cornea, is the result of an influence from the lens or possibly the edge of the optic cup; eye rudiments implanted under body-skin frequently cause the host skin over the implant lens to adopt this mode of development. Immediately outside the lens, cells become sparse, resulting in a fluid-filled **anterior chamber.** The **iris** develops from the edge of the optic cup, its circular muscles developing within the substance of this tissue (neural in origin). Melanin pigment is deposited in the iris (continuous with the layer of pigment in the outer wall of the optic cup, the **tapetum**) usually first on its inner aspect (the human iris looks blue or grey) then later, if genetically specified, on the outer aspect (brown or hazel human "eyes").

Behind the lens, in the cavity of the optic cup, the **vitreous body** is now secreted. There is doubt as to whether it forms around the **hyaline artery** (connecting the back surface of the lens to the retinal blood system) or is secreted from the retinal surface.

The eye is now more or less ready to function. The detailed interconnections of the nerve cells in the retina, and the apparatus (e.g. the **canal of Schlemm** draining the anterior chamber) for maintaining this function cannot be described here. The six **extrinsic eye muscles** are developed from the first three myotomes and are innervated by the first three ventral root nerves (Fig. 45). This "opportunism" in embryology is very common, as we shall see.

THE EAR

The ear, too, is a complex organ developmentally. Its initiation occurs at about the same time as that of the lens, with the appearance of an auditory placode on either side of the anterior part of the hindbrain. This is usually denoted by a thickening of the epidermis of this region (see p. 73), under the influence both of brain medially and of foregut roof more ventrally. The result of the sinking of this placode into the mesenchyme is a vesicle, but the details vary among the classes of vertebrates (Fig. 32). This is the **auditory vesicle** and is the primordium of the most important part of the ear, the **membranous labyrinth** and chambers. The dorsal part of this vesicle remains small and may retain a duct to the outside, the **endolymphatic duct,** especially in selachians.* The more ventral part of

* In adult elasmobranchs the duct is blind but nevertheless important.

the vesicle swells into an hour-glass shape, forming the **utriculus** antero-dorsally and the **sacculus** posteroventrally. The **basal papilla** projects as a medioventral nipple on the sacculus and the **semicircular canal** rudiments make their appearance as "lappets" on the utriculus (Fig. 50).

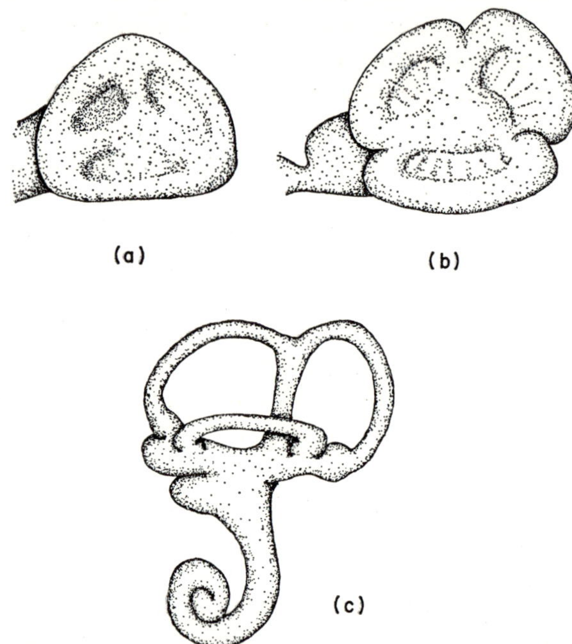

FIG. 50. *Development of the ear. (a) The auditory vesicle is becoming "shaped"; lappets have appeared in three planes. (b) The edges of the lappets have thickened. (c) The thickenings have become patent as the three semicircular canals and the coiled* processus cochlearis *is beginning to form the cochlea.*

Meanwhile, the dorsal root ganglia have been appearing above (and between) the myotomes along both sides of the spinal cord and brain. The third dorsal root lies just above and behind the auditory vesicles, nestling in the groove between utriculus and sacculus and extending dorsally. Most of it remains concerned with the sensory innervation of the hyoid segment and becomes the seventh (facial) nerve and ganglion, but the posterior part becomes more or less constricted off as the **eighth**

(auditory) ganglion. This now sends nerve processes into the epithelium of the forming sensory areas of the ear (the **maculae** and **cristae**) and nerve fibres into the brain, the **vestibular nerve** (and the **cochlear** in mammals and birds) which organize the part of the brain they invade so that the **nucleus angularis** and the **nucleus magnocellularis** form. Removal of the auditory placodes, vesicles or ganglia results in absence or underdevelopment of the brain nuclei of the operated side.

An important question at the time of writing this account concerns the association of the nerve endings in the sensory areas with the sensory hair cells they connect with. Do the nerve processes determine that a particular cell shall be sensory by contacting it, or do those cells which are intrinsically determined to become sensory "call" nerve endings to them? And how is the orientation of sensitivity determined? Each nerve cell of the ganglion connects with a number of sensory cells of similar orientation in each macula *and* each area in the macula, containing cells of a variety of orientations, is connected with a number of nerve cells. This provides for preliminary analysis of the information at peripheral level (as in the retinal nervous connections of the eye), but makes experimental analysis of the above questions very difficult.

Further development of the membranous labyrinths varies in different vertebrate groups; in mammals the basal papilla develops the coiled **cochlea,** and in most groups there is elaboration and subdivision of the sensory areas. The lappets on the utriculus flatten, obliterating their cavities except around the edges where they persist as tubes, forming the **semicircular canals** in their several planes (Fig. 50). (A useful analogy is of a thumb and finger grasping the edge of a lump of plasticine and squeezing: the centre becomes thinner and thinner, leaving a ring around the edge of a central hole attached to the rest of the plasticene block.) The central shelf usually breaks down completely, leaving the canals attached only by their ends; one end of each canal has a swelling, the **ampulla,** in which is situated the **crista** with its **sensory hair-cells.** The peculiar paths of the nerves to these cristae may result from their original paths over the lappets.

The first or hyoid branchial pouch lies immediately ventral to the developing labyrinth and, indeed, its dorsal wall is pressed against the ventral side of the sacculus/lagena. It also has approached the epidermis to make a standard gill-pouch (Fig. 43). In fishes, this usually breaks through and there is no special connection with the ear except that they are close to each other. In the land vertebrates, on the other hand, this ectoderm/endoderm junction forms the **eardrum** or **tympanum,** and the contiguity between the pouch and the membranous labyrinth remains soft as the **foramen ovale** and/or the **foramen rotunda** (the **round** and **oval windows**). The cavity of this pouch thus remains as the **middle ear cavity,** and drains and becomes air-filled from the pharynx when the animal starts to breathe air. The connection between it and the pharynx becomes restricted to a narrow passage, the **Eustachian tube.**

Meanwhile the various coatings have been deposited over the surfaces of the head organs, as previously described for the eye. The pia mater over the brain and the connective tissue coat over the membranous labyrinth appear at the same time. This provides support as a thick "basement membrane" for the epithelium of the labyrinth. The gut and skin also acquire a more tenuous "backing" at this time or a little later. The arachnoid coat does not sheath the inner ear, in the same way as it did not sheath the eye lens, but instead a fluid-filled space with sparse cells appears around the organ. The fluid is the early **perilymph** and is vitally important in the organization of the cochlea; the space containing it abuts onto the dorsal surface of the hyoid pouch, slightly separating this from the sacculus/lagena/cochlear process which was previously touching it. (The arachnoid only remains vascular over the surface of the cochlear process of the labyrinth, where it becomes the **tegmentum vasculosum;** the rest of the ear does not have the degree of vascularization of the eyeball.) The dura mater is now laid down over the brain and around the forming ear. It surrounds the whole organ, as with the eye, but does not intrude between the pouch and perilymph at the windows, or between the ganglion and the labyrinth (just as it did not intrude between iris and lens in the eye). It surrounds the middle ear to a variable extent, as well as the inner ear (where it later becomes the **bony labyrinth**), and forms a tube around the auditory ganglion and nerve right up to the brain. Through this tube, perilymph is continuous with the fluid surrounding the central nervous system *inside* the dura. It does intrude between the layers of the tympanum as this extends in area, giving it a tough middle layer except at the original point of contact of endoderm and ectoderm, which remains thin and may form a pore if the epithelia break down when they have air on both sides of them.

It often puzzles students that the earbones seem to be out in the free space of the middle ear and yet they are said to be homologous with gill bars. The situation is really very simply described. The first gill pouch, like the others, has gill bars which form in its posterior wall (*hyoid arch* in Fig. 46). The area of contact between this pouch and the skin grows as the middle ear develops and the middle-ear cavity swells, increasing the area of the tympanum. The bones of the arch soon appear as a bulge on its posterior wall and, as the cavity swells still further, it pushes around them and meets behind, leaving the arch as a pillar, the **columella auris,** surrounded by the endodermal lining of the middle ear and outside this by the cavity of the middle ear. The branchial blood vessel is also isolated in this pillar. This mode of formation accounts too for the strange path of the stapedial artery, around which the bone of the stapes is deposited.

Contributing to the ear of higher vertebrates, then, are the following: epidermis, as placode and tympanum; dorsal root ganglion; gill pouch endoderm; membranes comparable to pia, arachnoid and dura; and cartilage of gill bars (derived from neural crest) as well as bone and cartilage deposited as part of the general skull-deposition. All must be formed in appropriate order and position for a functional ear to result.

THE NASAL ORGAN

The development of this organ is deceptively simple to describe. In the jawed vertebrates, the appearance of the anterior end of the neural plate occurs in response to influences from the prechordal plate and the front end of the gut; these cause the anterior epidermis to thicken and form symmetrical **nasal placodes.** These dimple in and expand into vesicles whose rear walls are apposed to the front of the brain, which has by now been overgrown by epidermis and sealed off, except for the neuropore. Sensory nerve cells now appear in the epithelium of the placodes and send processes back into the brain, where in response to this invasion the **olfactory lobes** appear and grow. The source of these sensory cells is, however, quite uncertain; they may be transformed cells of the nasal epithelium, they may be cells of the brain which cross over into the epithelium when the two are apposed, or they may be cells comparable with neural crest which are "trapped" over the anterior end and form a kind of median dorsal root ganglion. It is difficult to decide between these alternatives by extirpation or transplantation, because this region is very susceptible to operation damage and most interference results in developmental peculiarities explainable by any of the above suggestions.

These nerves do not seem to make many cross-connections on their way into the brain and it seems, therefore, that integration of the incoming signals must take place in the circuitry of the brain tissue itself, rather than peripherally as in the ear and eye. (They do, however, form tight bundles of a few fibres, and "cross-talk" may occur in the olfactory lobes.)

Between patches of sensory cells the epithelium becomes ciliated and mucus-secreting and is usually thrown into complex folds. Again the three layers comparable to pia, arachnoid, and dura ensheath the organs and the dura forms the **nasal capsules** and perhaps the **turbinals.**

Although the above simplified description applies to all vertebrates, it should be emphasized that aquatic vertebrates use the olfactory organs for a sense which is probably much more like our sense of taste than our conception of smell; but, even in terrestrial vertebrates, taste and smell are really very closely related.

THE ENDOCRINE GLANDS

The **pituitary gland** arises from a stomodeal pocket, the **hypophysis,** which forms the **anterior lobe.** Some neural crest and head mesenchyme form the **intermediate lobe.** A downpushing from the forebrain, the **infundibulum,** forms the **posterior lobe** (Plate XIV).

The **thyroid gland** arises as a groove in the floor of the pharynx which loses its connection with the gut floor and whose walls become glandular. It usually moves posteriorly during development.

The **parathyroids** derive from supra-branchial bodies, as does the **thymus.**

The **Islets of Langerhans** are special parts of the original endodermal pancreatic outpushing which lose their connection with the duct.

The **adrenal glands** have a dual origin. The **cortex** derives from mesodermal nephrotome, while the **medulla** comes from neural crest tissue similar to that of the sympathetic ganglia.

The gonads derive from two sources. The germ cells derive from cells which lie in the undifferentiated "endoderm" and migrate to the gonotomes; the interstitial tissue is nephric in origin and is the prime endocrine source.

THE URINOGENITAL SYSTEM

We will first consider the condition in the frog and then that in the mammal.

The rudiment of the excretory system in the tadpole arises well before hatching as a longitudinal thickening of the somatic mesoderm on each side of the coelom, below the myotomes. This mesoderm is given the general name of the **intermediate cell mass,** but in many animals it has differentiated into nephrotomes, one from each somite (p. 110). Because of influences from the gill pouches and the hindbrain its front few segments (roughly segments 2 to 8) give rise to the **pronephros,** which consists of about three twisted tubules each opening into the coelom by a funnel (**nephrocoelostome**). Somewhere within these tubules, usually near the mouths of the funnels, a sacculated outgrowth of splanchnic layer appears. It is known as the **glomus** and becomes filled with blood from the dorsal aorta. Due to blood pressure and the thin wall of the glomus, constituents from the blood can pass into the lumen of the pronephric tubes. Meanwhile, the inner blind endings of these tubes, which in fact lie in the outer edge of the nephrotome, join together to form a longitudinal tube, the **segmental** or **archinephric** duct. This pair of ducts (one on each side of the body cavity) now grow backwards. As they pass from segment to segment in a posterior direction, they induce the formation of the functional adult kidney in the **mesonephric** region (roughly from trunk segments 10 to 20). By the time of hatching, they have continued posteriorly in the dorsal wall of the peritoneal cavity, looped around its posterior end and reached the cloaca by 'climbing up' the proctodeum (p. 105).

The frog kidney, **mesonephros,** consists of a series of paired masses of cells in the nephrotomes in these middle segments, each of which develops into one of the **kidney tubules,** having at one end an opening into the segmental duct (now called the **mesonephric** or **Wolffian** duct) and at the other a **Malpighian capsule** (secondary nephrocoel) with a **glomerulus** (Fig. 51). The testis takes over the most anterior of these mesonephric tubes and uses them to form the **vasa efferentia,** while the more posterior tubes function as the adult kidney. At maturity, therefore, the Wolffian duct functions to transport both sperm and urine in male amphibians. Just before metamorphosis the pronephros and the front part of the segmental duct degenerate.

The **oviduct** now arises as a new structure called the **Müllerian duct.** It is formed as a longitudinal tract of the peritoneal epithelium outside the kidneys which becomes converted into a canal, the front part by being grooved and then closing in, the hinder part by hollowing out. The anterior end of the groove does not close, but remains as the internal opening of the oviduct. Thus in female amphibians the eggs do not share a common duct with the urine, as the sperms do.

The **gonads** are formed in thickenings of the coelomic epithelium (gonotome) ventral to mesonephros one on either side of the mesentery, on the dorsal wall of the peritoneal cavity. No distinction between the sexes can be seen until metamorphosis takes place, since both Wolffian and Müllerian ducts are present in the larvae. In the male, the Müllerian duct degenerates at the late tadpole stage, leaving only a minute vestige at metamorphosis; the anterior part of the Wolffian duct, which should attach to the testis, degenerates at this time in female tadpoles.

In the frog, and in amphibians generally, the segmental duct tends to lose its organizing capacity after it has traversed about two-thirds of its journey to the cloaca; no more mesonephros results behind this as the tube passes under or through the remaining posterior nephrotomes. But in most reptiles and in the apodan amphibians, kidney tubules appear all the way along the path of the segmental duct. This situation occurs also in many fishes, especially the selachians. Where the front end of the mesonephros is used in the male for genital functions, as in the frog, this defines the mesonephric area and posterior to this, into the back end of the body cavity, the kidney organized directly by the segmental duct is called **opisthonephros.**

The functional kidney of mammals, birds and reptiles is called **metanephric** and its development is more complex. The pronephric rudiments appear (probably under the inductive influence of the hindbrain, the gills, and perhaps blood vessels of the transverse septum), and unite to form the pronephric duct, just as in the frog. This pushes posteriorly, organizing mesonephric tubules which join to it. Thus, as in the frog, the mesonephric or Wolffian duct is formed. It runs posteriorly but, again as in the frog, it does *not* organize opisthonephros in the hind region. It then bursts into the ventral side of the cloaca, probably in the endodermal part, having looped around the rear end of the peritoneum. A small diverticulum from the duct, near this junction with the cloaca, then grows up around the wall of the body cavity and its tip, upon meeting unorganized nephrotome in the posterior end of the animal (in the opisthonephric region in fact), induces this to become **metanephros,** the functional kidney of adult mammals and birds (Fig. 51). Only one or two nephrotomes are involved (roughly trunk segments 31 to 33), but the tubules branch and subdivide to give the many glomeruli and tubules

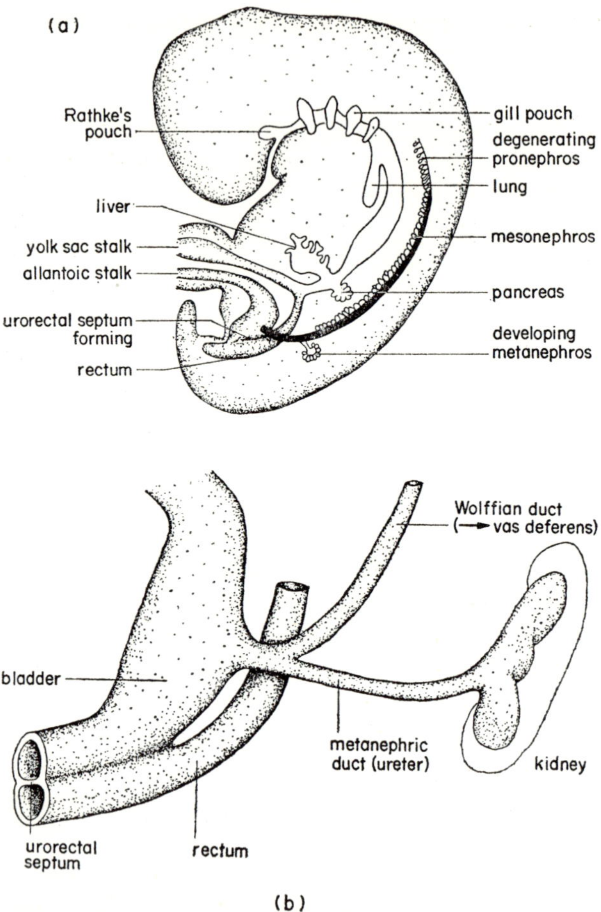

FIG. 51. *Development of the urinogenital system of the mammal.* (*a*) *An early mammal embryo.* (*b*) *Enlargement of the posterior organs at a slightly later stage.*

characteristic of the adult organ. The new duct is now the **metanephric** duct or **ureter.** Due to the growth of the cloaca, the bottom end of the Wolffian and metanephric ducts get "sucked in" to the wall of the cloaca; their openings therefore separate, and their walls form part of the wall of the urogenital sinus (see next page).

It should be recalled that in the mammals the hindgut changes from an **O** to an 8 shape in section, due to the development of a shelf of tissue pushing in from the sides of the gut, separating a dorsal rectum from a ventral urogenital sinus (p. 105). This shelf (the **urorectal septum**) becomes continuous with the front wall of the allantoic diverticulum (Fig. 51), so the cavity of the **urogenital sinus** is continuous with the cavity of the allantois (and the cavity of the rectum is continuous with the midgut cavity and yolk-sac). The ureter and the Wolffian duct, which originally opened into the urogenital sinus by one opening have had some of their length pulled in to form part of the wall of the urogenital sinus and their openings are now separate: the ureter now opens anteriorly into what will be the **bladder;** the Wolffian duct opens more posteriorly into the **urethra** (Fig. 52). (In man anterior/posterior are confused here because of his upright stance; posterior = ventral!).

The Müllerian ducts have, meanwhile, begun to form and lie along the dorso-lateral wall of the peritoneal cavity; their posterior ends approach each other and enter the urorectal septum mesoderm, now called the **broad ligament.** Here, they fuse and proceed down the septum between rectum (dorsally) and urogenital sinus (ventrally). The **uterus** is formed in the broad ligament and the **vagina** develops from the junction of Müllerian and urogenital sinus systems. The situation is possibly here further complicated by the presence of the old primitive streak on the cloacal membrane and material from kidney ducts on the wall of the urogenital sinus; this makes the derivation of vagina and the various accessory structures difficult to establish.

In the female, therefore, the bladder opens (by the urethra) separately from the vagina, which is more dorsal. In many mammals an ectodermal pocket, the **vestibule,** may develop in the urogenital region (e.g. in the human). It is often extremely deep (as in elephants and many rodents) and appears to merge into the vagina, so that the urethra opens internally into it on what is apparently the anterior wall of the vagina (e.g. the mouse). The original **clitoris** is located deep inside the female opening in these cases, and a new erectile structure is developed at the entrance to the vestibule (again as in the mouse).

The ovaries have developed near the proximal ends of the Müllerian

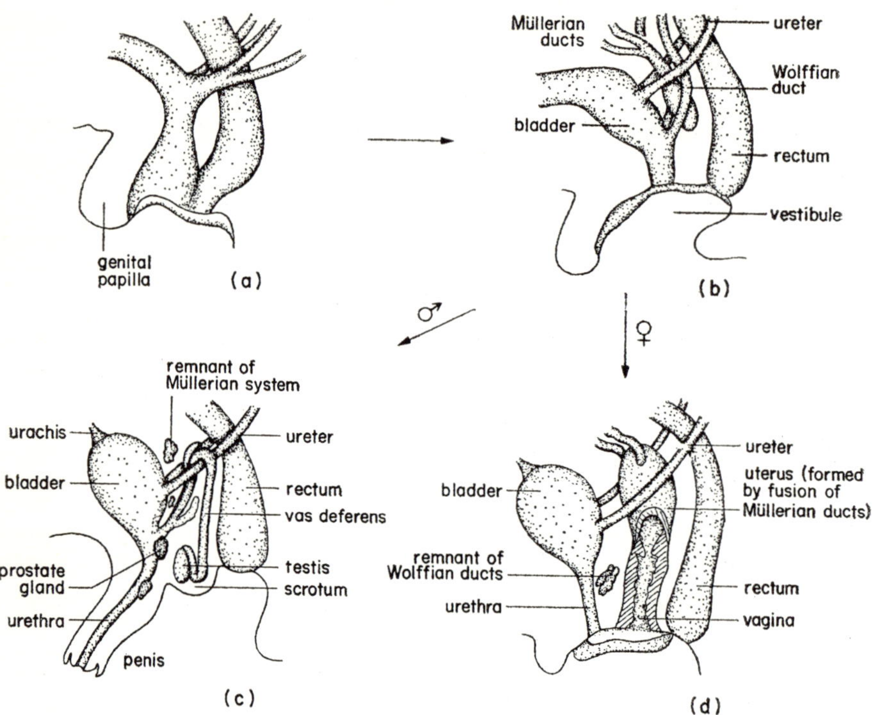

FIG. 52. *Later development of the mammalian urinogenital system. (a) Early stage before separation of the Wolffian and metanephric ducts (compare with Fig. 51 (b)). (b) Slightly later than (a), showing the common intermediate stage in which Müllerian and Wolffian ducts are present. (c) Development of the male system and consequent regression of Müllerian ducts. (d) Development of the female, with regression of the Wolffian ducts.*

ducts, now called **Fallopian tubes.** They are usually more or less cut off from the genital and nephric ridges (gonotome and nephrotome) and have their own little "mesenteries". These, the **mesovaria,** are continuous with the broad ligament, leaving the ovary intimately associated with the fallopian tube and effectively detached from body wall. The mesonephros and Wolffian ducts degenerate in the female, sometimes leaving little cysts around the mesovaria and the fallopian funnels.

In the male, the mesonephros forms connectives with the developing testis and becomes the **epididymis,** its tubules are the **vasa efferentia** and the Wolffian duct the **vas deferens.**

Originally the Wolffian duct followed the dorsal wall of the peritoneal cavity, against the external edge of the nephrotomes but, of course, outside the peritoneal membrane (that is, not in the peritoneal cavity) and curved right round the posterior wall until it reached the cloaca, ventrally. In the opisthonephric region it does not organize the nephrotomes in the amniotes (though, of course, two or three of these nephrotomes *are* later organized into the metanephros by the branch from the Wolffian duct, the ureter). As the animal grows, the Wolffian duct stays relatively short and as a result pulls out a web of the peritoneal fold which forms a "mesentery" for it, running from dorsal to ventral on each side of the posterior end of the body cavity. This fold, called the **gubernaculum,** thickens and acquires much connective tissue and muscle fibres. It contracts and draws the epididymis and testis down into the most posterior, ventral part of the ventral cavity, where **scrotal sacs** have formed ready to receive them. Because of this descent the vas deferens has been looped over the ureter. The testis retains a connection, the **anterior suspensory ligament,** with its original position, anterior to the kidney, in which run the **spermatic artery** and **vein;** where the vas deferens runs with these vessels up out of the scrotum the system is called the **spermatic cord.** (In marsupials, however, the process occurs slightly differently and the scrotal sacs are in front of the penis.) The urethral opening in the male is progressively enclosed and comes to open at the tip of the **penis.** Thus, in the male the functional urinary and genital openings retain their connections with urogenital sinus; but note that sperm and urine only share the very posterior end of the system, unlike the situation in animals with meso- or opisthonephric kidneys. Most of the Müllerian system remains vestigial in male vertebrates.

Let us now consider the vascular relations of the kidneys, with special reference to the frog. The pronephric and mesonephric tubules arise below the posterior cardinal sinus and bulge into it, almost occluding it. Anterior to the mesonephros a new sinus, the **posterior vena cava,** appears which may follow the path of one or other posterior cardinal sinus or run ventral to the dorsal aorta in the dorsal mesentery. This takes the "short cut" to the sinus venosus, dropping into the dorsal mesentery of the gut anteriorly over the fused mesogastria (Fig. 44) and passing through liver substance which has invaded mesenteries anterior to the stomach. The remains of old anterior and posterior cardinals persist, from posterior to anterior, as the **renal portal veins** behind the kidneys; part of the **inferior (posterior) vena cava** from the kidneys through the liver; the **azygos vein,** parts of the **subclavian veins,** and the **jugulars;** these last derive from parts of common and anterior cardinal sinuses. The male mammal cannot still use the mesonephros as a functional kidney, because it has taken over a genital function. The posterior vena cava extends back to just in front of the *metanephric* kidneys in these animals and so by-passes the old posterior cardinal for almost its whole length. This avoids the necessity for the blood from tail and legs passing through the epididymis. Later, the kidneys receive an artery and the posterior vena cava by-passes them too, only remaining connected by the **renal vein.**

GENES AND DEVELOPMENT

We have already seen (p. 27) that the nuclei of early embryos are virtually passengers in the cytoplasm of the blastomeres, except in mammals (p. 83) Development of nearly all animals is determined up to the phyletic stage (p. 68), by the spatial organization of the egg cytoplasm and its m-RNA, proteins and other important molecules which were stockpiled during oogenesis. These early embryological processes are usually fuelled by food and energy reserves which have also been passed from the mother's surplus (p. 12).

Very little is known in detail of the ways in which maternal genes act to give the egg its spatial organization, or how the m-RNA transcribed during oogenesis is packaged and labelled "open at 12 hr" or "open in animal-derived cells of late blastula". But the demonstrable existence of this organization, and of the packaged m-RNA, means that very different questions are being asked now about the differentiation of the characteristic cells of metazoans from those asked 15 years ago. Liver, kidney, or skin cells all have nuclei with the same genes; yet they all exhibit different forms and functions, in terms not only of the major proteins secreted or contained but also in their shapes, sizes, division rates, susceptibility to radiation or cell poisons. This **differentiation** of these cells is now seen as a continuation and further development of differences in the cytoplasm from which they were cleaved. It used to be thought that, because blastomeres could be forced to be totipotent, there were no differences between them. Theories of differentiation used to start by assuming that cells were initially similar and attempt to account for later differences. Now we accept that many of the initial *cytoplasms* were different and will become more different after cleavage has separated them; thus the answers to differentiation lie in *cytoplasmic control* of nuclear function rather than in differences between the nuclei themselves.

The evidence for cytoplasmic control of nuclear differentiation is very clear and comes from many kinds of experiments, of which only two will be mentioned. Gurdon has shown that "differentiated" endoderm (gut) cells of a late tadpole, when inserted into an enucleate frog's egg, can often behave just like the eggs' own nucleus and produce a normal frog. All of this frog's tissues have nuclei whose ancestor was the transplanted gut nucleus, which has proved its capability to contribute to muscle, nerve, skin and so forth. The cytoplasm of the egg "took it back to the beginning", as it were; indeed, nuclei from spinal cord neurones (which would normally never divide again) have been put into frog oocytes, where they have undergone meiosis with the resident nucleus and even contributed to embryonic development. It is *possible* (see below) to explain these results by assuming that all tissues have a few multipotent, continually embryonic stem cells, whose nuclei are the only ones which can contribute in the way described (and assuming that the nuclei of "really" differentiated cells contribute to the failures of such experiments). However, this explanation seems very unlikely; firstly, because success rate in transplant experiments is too high, exceeding 20 per cent; secondly, because many other experiments suggest that cytoplasm is the important factor in the resulting development. Harris has used killed virus particles to fuse together cells of different tissues; indeed from different *species*. In the resulting binucleate cell the originally less active nucleus, e.g. that from a chicken red blood cell, whose nucleus has neither RNA nor DNA activity, becomes as active as the more active nucleus, e.g. that of a mouse lymphocyte making m-RNA for antibodies—in this example the chicken nucleus starts to make the m-RNA for chicken haemoglobins again. Because the nuclei are not actually in contact with each other, the influence on the chicken nucleus must be at least via the cytoplasm and the general belief is that, taken together with such experiments as Gurdon's, the cytoplasm has been demonstrated to be "boss". As a further example, if a cancerous cell (e.g. a Hela cell) is combined with a cell with an inactive nucleus, the originally inactive nucleus undergoes rapid mitotic divisions and also secretes m-RNA in the interphases.

There are many other cases in which the cytoplasmic condition can be seen to regulate nuclear activity. In insects, the first nuclear divisions result in some tens of nuclei in the uncleaved cytoplasm (p. 52), which move to the surface apparently at random. Only those which arrive in the pole plasm will later become germ cells; fruit-flies homozygous for *gc*-(grandchildlessness) make eggs without pole plasm, which produce normal but sterile progeny. If pole plasm from a normal egg is injected into a *gc*-egg, that egg can then produce a fertile fly. The other insect nuclei soon cease dividing in most of the larval tissues, but continued endoreplication of the DNA gives **giant chromosomes.** In the different tissues, at different stages, entirely predictable constellations of genes can be seen to be active in these giant chromosomes. The active regions make a bulge or **'puff'** (or **Balbiani ring**) on the chromosomal thread, which can be demonstrated to be making m-RNA. Certain "puffs" are specific to certain cells, whereas other "puffs" will appear in many kinds of cell. The same patterns of puff formation and presumably therefore of gene activity can be induced experimentally by simple changes in Na/K ratio or pH, as well as by combinations of moulting and juvenile hormones, all of which presumably work via the cells' cytoplasm or membranes.

Biochemical studies of the RNA produced by a variety of tissues, mostly bird or mammal, have demonstrated that many cells share certain kinds of m-RNA production, presumably coding for "housekeeping" enzymes and other proteins which are common to many cell types. Others, for example red blood cells or silk-secreting cells, have a unique m-RNA for their unique protein product. Some, like muscle cells, produce large amounts of m-RNAs (e.g. for actin and myosin) which very many cells produce but in small quantities.

The acquisition of differentiation *may* be sudden, but is usually progressive. Let us consider the fate of the grey crescent of the frog's egg. This cytoplasm normally becomes incorporated into the gut of the embryo. Some of it forms the cytoplasm of the germ-cells, which wander up the dorsal mesentery to their final positions in the dorsal wall of the body cavity. Originally, in the just-fertilized egg, the grey crescent material can in fact contribute to much more than gut or germ-cells. If, at the 4-blastomere stage, the two anterior blastomeres are separated from the posterior blastomeres (which each contain half of the original grey crescent), then the anterior pair usually do not develop further than an aberrant blastula; but the posterior pair form a complete, but small, embryo. The grey crescent material in this case has been shown to contribute to notochord, to lateral mesoderm and the anterior somites. The final resting place of the grey crescent cytoplasm in the normal case, the gut wall and the germ cells, cannot in any circumstances (as far as we know) be persuaded to become notochord or somite tissue. This experiment, as well as those involving growth of different adult tissues in tissue culture (see below), shows the progressive decrease in the potentiality for diverse development, which is characteristic of progressive cell differentiation.

However, in many tissues in the adult animal there may still be considerable ability to choose among several alternative forms. A good example is found in the urinogenital systems of mammals, where many different tissues show cyclical changes related to hormone concentrations. The cells of the uterine epithelium, for example, may be tall columnar and mucus-secreting or very inactive and flat at different phases. The epithelium of the vaginal wall in the mouse may form a low cuboid epithelium, or may contribute to squamous epithelium of many cells in thickness. Cells of the epidermis of mammals, if they come under the appropriate inductive influence of a dermal papilla, are capable of undergoing the many complex changes associated with their production of, and incorporation into, a hair follicle. These various possibilities which remain open to the individual cells of adult tissues have been called **modulations** to distinguish them from the apparently permanent differentiation of many other tissues.

Just how permanent differentiation is has been the subject of great controversy. Two lines of evidence have been thought to show that differentiated cells may de-differentiate. When adult cells are transplanted to tissue culture they almost always lose their characteristic shapes (Plate XIIIa) and they may lose their characteristic physiology as well. However, all experiments in which such cells have been replaced in an animal have demonstrated that they *still* cannot form anything but the kind of tissue from which they originally derived. Some "cell lines" may have become cancerous, but the tumours they produce when injected into a host are still of that family characteristic of the original tissue. Therefore, the *apparent* de-differentiation in tissue culture is to be regarded as a *modulation* directed by the new conditions in which the cells find themselves and not as a turning back along the embryonic path.

The other case in which de-differentiation has been postulated is that of **regeneration.** If a newt loses its tail or one of its limbs, the wound shows a very odd kind of behaviour. A pile of cells accumulate over the gap (the **blastema**) and a series of movements and differentiations is initiated, controlled at first by local influence, later by the action of ingrowing nerves; a new, perfect limb is produced. Obviously, cells and tissues of all the kinds found in the digits were not present in the stump; the cells of the stump have, nevertheless, produced a normal hand. It is nonetheless very doubtful whether any of the cells which contribute to the blastema and later to the perfect limb, have changed from the kind of tissue of which they were originally a part. Connective tissue cells have for a period "gone into disguise", but have probably only formed connective tissue and muscle; similarly, epidermal cells have probably only formed skin. A complicating factor here is the **interstitial cell** story. In coelenterates, it has long been supposed that the tiny cells, found close to the mesogloea in the ectoderm and gastroderm, are replacement cells and can give rise to any of the kinds of cells which characterize the animal.

For example, they are said to form the germ cells and to provide a constant supply of nematocysts. Some students of regeneration have gone further to suggest that all the tissues of the metazoan body contain similar nests of **stem cells.** These would be cells which are embryonic in nature and have not yet decided to differentiate in any particular way. According to this theory, these are the cells which form the amphibian blastema and which are then differentiated progressively as the limb is formed. (They *could* also be the only cells whose nuclei could contribute to a whole new embryo, see p. 137.) Unfortunately the evidence from the enormous number of damage and regeneration experiments which have been performed does not tell us surely whether such cells exist, or even indeed whether the famous interstitial cells of *Hydra* perform this function.

It cannot be emphasized too strongly that cellular differentiation is not merely a change in form or structure. It involves far-reaching changes in the whole physiology of the cells concerned and these cells may acquire a whole range of new responses to chemical and physical stimuli. They may become able to secrete new substances or to perform functions (like the contraction of muscle or the impulse conduction of nerve) which were impossible to their ancestors. A cell system which has been much studied in this respect is the pigment cell system of vertebrates, especially that of birds and mammals. Its cells originate in the neural crest, from whence they migrate to all tissues of the body. (These neural crest cells are, by the way, the most likely candidates for the "stem-cell" which we have just been discussing.) That part of the neural crest cell population (melanoblasts) which will make pigment cells, behaves differently in the various tissues in which it finds itself. It is extraordinarily responsive to local influence and in the skin shows that its differentiation is controlled by factors both internal and external to the individual melanoblast. Some of these cells become stellate with long processes which are in close relation with epidermal cells (Plate XIII d, e, f); but they still may not make pigment to pass into these epidermal cells without a further cue from the epidermal cells themselves.

When pigment has been produced, even its colour may depend upon the kind of epidermal cells giving the cue. As an example let us consider the pigment cells of feather germs. Even in a brightly coloured bird like the Brown Leghorn cockerel, all the *melanoblasts* are the same in all respects. However, in the wing pinion feathers their progeny, melanocytes, produce massive quantities of black pigment which is donated to the epidermal cells which will become keratinized as they form feather. These feathers are therefore black. In the saddle feathers a different situation occurs. Identical melanoblasts receive a *different* cue and produce brown melanocytes which donate pigment. In the neck feathers the situation is different again. In the "dorsal" part of each feather cylinder, black pigment is produced and in the "ventral" part (which will be the edge of the feather) a pale yellow pigment is produced. Those pigment cells which find themselves in the rachis (central stem) of the feather often produce pigment of various kinds, but then seem unable to donate it. All these different responses of the Brown Leghorn melanoblasts are true differentiations and not simple modulations. This must be so because the pigment-producing cell, the melanocyte, rarely divides and new ones are produced from the melanoblast population. It must be realized that the melanoblast population occurs in all the tissues of the bird but that in most of them, for example liver and gut, no pigment is normally produced. The melanoblasts therefore respond in a whole variety of ways, ranging from the production of no pigment at all through the production of a little yellow pigment to the production of copious black pigment, in response to a range of local tissue factors. These melanoblasts are admittedly an extreme case; this is associated with their extensive distribution in the tissues and the considerable evolutionary importance of their function, which is the production of colour. Unfortunately, although much is known of the biochemistry of pigment formation, nothing has yet been discovered about the inductive processes concerned.

In fact, the big gap in contemporary embryology is a coherent theory of differentiation. Many theories account for particular cases, especially of a biochemical nature, in terms of competition between enzyme systems which may result in the domination of the cell by one particular system. This does not, however, account for the structural changes which are at least as important for the full function of the fully differentiated cell.

Recent emphasis has been on theories which involve a certain subclass of the surface molecules of differentiating cells. At least in mammals, these molecules are evolutionarily closely related both to antibodies and to those specific antigens on tissues which enable recognition of "foreignness" (and so permit graft rejection). The semolecules are all dimers (doublets) or tetramers of repeated sequences, the amino-acid sequence of which is like a common blood protein, β_2-microglobulin. They occur on cell surfaces together with β_2-microglobulin itself. There is some evidence that the differentiations of male germ cells in several mammals are caused by the linking of one dimer of this kind, the so-called Y-antigen, to this type of surface site: this attaches to cells of the developing gonad and *makes* them become testis. On the other hand, undifferentiated cells, e.g. cells of the blastocyst of mammals, show a quite different spectrum of these supposedly differentiation-producing molecules and some tumours have been shown to produce a "blastocyst-cell-surface-like" spectrum. Ohno has suggested that this is a general situation and that these sites are surface triggers for different kinds of *nuclear* differentiation, perhaps via different kinds of molecules in different tissues. Virally-infected cells have viral proteins at these sites and are made to perform viral replication even when no DNA-replication would be expected in this kind of cell; some viruses perhaps cause tumours in this way, by turning off the cell's differentiation and turning on DNA-replication. Ohno has further suggested that the whole tissue-and-individual-specificity system, with specific antigens and antibodies polymorphic in each species, is the basic metazoan defence system against viral attack. Cells whose differentiative pathways have been perverted to viral replication can be recognized by antibodies because they *have* viral protein on their surfaces; thus they can be destroyed. This is referred to as **immune surveillance** and could be the body's main defence against cancer, the perversions of differentiation. Because all individuals of a species have *different* cell antigens the virus cannot mutate to mimic the natural antigens, because these vary from individual to individual as do the antibodies which recognize them. This, according to Ohno, is why different histocompatibility types (for example HLA in man) are susceptible or resistant to different viral diseases (e.g. in man, allele 27 at the B-locus is associated with ankylosing spondylitis).

LIFE CYCLES

Embryology is part of the process of reproduction. Many of the varied reproductive strategies of animals are associated with special embryologies. For example, the phyletic stage, during which control of development is being passed from the maternal to zygote genome, is a way of using the mother's complex, already-tested physiology and morphology to give the embryo a basic structure, a framework within which to develop; it does not take its own direction right from fertilization. Other maternal effects on early embryology are many (p. 12), ranging from differences in quantity of yolk to nicotine or thalidomide in the uterine environment! By no means all of the maternal effects on embryonic development come through the DNA or even the maternal m-RNA of the oocyte, but they are nevertheless part of the inheritance of an embryo, *if* they result from that particular "choice of parent" (p. 13).

We have already considered *r* and *K* strategies briefly (p. 14). An important consequence of the phyletic stage for some embryos and larvae should be considered here. Some larvae, notably the trochophore and dipleurula mentioned in this text (pp. 45 and 56), *are* really the phyletic stages. So the zygote genome has had little effect on their structure yet; they have used the "guaranteed" maternal programme (and material) to get to this stage. Even though the species may be an *r*-strategist and the embryos may be genetically very diverse, they have had a privileged early development—so far. It is common experience that most polychaets die as trochophores, and so on. Many other larvae, even those of insects and amphibia, get a considerable start in their early development in both material and informational terms, from which their own genome can take over. In *Xenopus*, tadpoles which lack nucleoli and are thus unable to make any of their own ribosomal RNA, the oocyte (maternal) contribution of ribosomes still permits development to late tail-bud stage in most tadpoles.

The larva, even if not the phyletic stage, is considerably specialized for its *own* way of life and there are many obvious advantages to be gained by avoiding competition with its parents. However, this means that the change to the adult structure must often be via a **metamorphosis.** Polychaet worms, many insects, most echinoderms and frogs have dramatic metamorphoses.

The metamorphosis of the trochophore into the worm has been described (p. 46), and it will be recalled that almost all of the substance of the adult worm is derived from only a small part of the larva. This is not an uncommon situation. Much of the substance of an adult insect is derived from the **imaginal discs** of the pupa, which start the major part of their development in the absence of juvenile hormone. Echinoderm larvae often form the adult from a relatively small and usually highly asymmetrical part of the larva, the rest of the larva being reduced to the status of a degenerating appendage in the more extreme cases. Another special case of metamorphosis is that of an Order of Crustacea, the Rhizocephala, of which *Sacculina* parasitizing crabs is a common example. The adult resembles nothing so much as a mass of fungal hyphae spreading through the tissues of the crab, with a **"fruiting body"** projecting under the abdomen as a cream-coloured solid mass. The egg hatches into an ordinary crustacean larva, a nauplius, and this swims actively in the sea by means of its appendages until it metamorphoses into a **cypris larva** like those of other barnacles. This cypris larva settles on a crab instead of a rock and moults to become a **kentrogon.** This cements its head to the carapace of the crab and a tiny mass of tissue passes from the base of the antenna into the crab's body. *This* develops into the adult, all the rest of the larva serving no further purpose. Note the relevance of this to an embryological dictum, attributed to Samuel Butler: to the embryologist there is no question as to which came first, the chicken or the egg—the chicken is obviously the egg's device to make another egg! It does not matter that the egg of *Sacculina* produces an adult which resembles a fungus more than a normal crustacean provided that more eggs result, which in their turn produce more adults.

Polyembryony, when one egg produces two or more embryos, occurs in many animals, of which the armadillo is probably the best known. A special category of this is **larval multiplication,** shown very dramatically in many parasitic flatworms, e.g. the liver fluke. Here each egg produces one **miracidium** which becomes a **sporocyst** in a snail; each sporocyst produces many **rediae,** each of which produces many secondary rediae; finally thousands of **cercariae** are produced from these. All of these potential adults result from one egg. Many coelenterates, e.g. *Obelia*, also show a comparable situation; each fertilized egg produces a **planula** larva which settles and becomes a **hydroid** or **polyp;** this buds and some products become **blastostyles** which each produce many **medusae;** the medusae represent the adult sexual form and hundreds of thousands may be produced from each egg.

Animals usually increase in size until they become sexually mature, when general body growth usually slows considerably or stops altogether. During this phase of growth it is very rare for the parts of the body to maintain their size-relationships with one another, to remain **isometric;** it is more usual for the proportions to change as the animal grows (Figs. 18, 27 and 38). This change of proportion is normally not uniform but changes with the absolute size of the animal, and is **allometric.** The relative sizes are expressed mathematically as $y = bx^{a}$ where $x =$ size of whole animal, y is the size of the organ or part under consideration, b represents size of organ for unit body size, and ^{a} is the **allometric constant.** This form of equation accurately describes the changing proportions of various systems during embryonic and larval life and may also be used to compare related species of different sizes. However, at various stages the allometric constant may suddenly change, resulting in the acquisition of a new body form. These changes are one of the ways in which genes determine animal form and, as might be expected, show relative heterochrony in different animals. Hormones from the gonads and other endocrine glands may also change the allometric pattern and result in typical "adult" characters, for example, the long fins of some male fish and antlers of deer.

EMBRYOS AND EVOLUTION

Animal evolution has produced and is producing a great variety of animal forms. The ways in which this variation appears in the life history of the animals concerned is part of the province of embryology. It will be realized that, in general, the earlier in development an innovation appears the more profound the effect it will have upon the adult form. As an example, a modification of gastrulation processes might be expected to change the course of development considerably; whereas modification of almost adult developmental patterns would have much less effect.

We have evidence from many sources that those changes which result in new animal forms may occur at any stage in the life history; we will now consider examples of such changes and their importance.

Firstly, let us consider caenogenetic modifications. These are those variations which benefit the young form, embryo or larva, and have no particular significance for the adult (except that they enable the adult to be produced!). Very often they are lost at hatching, birth or metamorphosis. Examples are the cilia of the trochophore larva and the embryonic membranes of vertebrates. Caenogenesis often allows larvae to exploit quite different environments from the adults of the same species; examples are the glochidium larva of *Anodonta* (the swan mussel) which is parasitic on fishes, or the numerous planktonic larvae of sessile marine forms. Selection then acts to fit the larval stages to this environment, so that metamorphosis to the adult becomes more and more profound. Many of the new larval characters now become available to the adult, if metamorphosis does not eliminate them. Examples of the use by adult forms of larval (caenogenetic) adaptations are the retention of the cranial flexure in man, or of torsion in gastropod molluscs. (This latter is in many ways an embarrassment to the adult and many species untwist at metamorphosis.)

As selection may operate in different ways upon larvae and adults it is not surprising to find that in some cases the last stages of the life history have been greatly shortened, suppressed, or indeed lost altogether. Such a situation occurs in many insects with aquatic larvae like May-fly. In the case of May-flies, the larvae live for years in the water feeding and growing, while the adult have no functional digestive or excretory systems and only live long enough to mate and lay eggs.

The common eel is probably another case of the same kind. Such a series has reached completion in many neotenous Amphibia (e.g. axolotl and mud-puppy) where the terrestrial stage has been lost altogether and the animals breed while in a form comparable with the larvae of their relatives. That the same has not occurred in the insects is probably due to the close interrelation in these forms between the endocrine stimulus for the change to the imago and for the maturation of the gonads; in amphibians, these are controlled by different hormone systems.

The converse situation, **paedomorphosis,** is the early acquisition of the adult state while the organism still has the larval modifications and general body form. The insects are probably derived by paedomorphosis from a form resembling modern millipedes, which have a six-legged stage in their embryology.

The vertebrates may have been derived from echinoderm-like ancestors by retention of their *larval* characters. In many ways the primates, and especially the anthropoid apes and man, may be said to be neotenous mammals. Here the juvenile mammalian characters (including curiosity) are retained into early adulthood. It will be seen that either neoteny or paedomorphosis allow caenogenetic modifications to become true adult characters and so may change the entire morphology of the descendants of the animals concerned.

Deviation, on the other hand, describes those changes which are adaptations, primarily by adults, to meet adult requirements. They normally result only in variety or species differences, but may be "pushed back" (via heterochrony) into the embryology of the animal and so cause progressively greater morphological changes. An example of this is the reduction in limbs found in some lizards, especially the skinks, which reaches a culmination in the slow-worm *Anguis*.

The processes of development cannot, of course, be modified indefinitely. Only some modes of development produce viable products. Therefore, such changes as do occur tend to be conservative and to adhere fairly closely to the old pattern. There are many modifications, few innovations. The mammal still has a vascular supply to its yolk-sac, despite the absence of yolk in it; without this, the hepatic portal system could not later be constructed. Similarly, the gill vessels of mammals resemble those of the *embryos* of our fish ancestors. An interesting **vestigial organ** is the pronephros; this is probably only functional in two adult vertebrates (marine fishes *Fierasfer* and *Gobiesox*), but its remains are present in all. During development, it serves to induce mesonephros and then this tubal system induces metanephros. Without the pronephros, another system must arise either to induce true kidney or to function as kidney. Such a major innovation only rarely occurs and the pronephros, although non-functional physiologically, is still retained in all vertebrate embryos.

Because of the conservative nature of developmental changes, the early stages of two related animals usually resemble one another more than later stages. This, von Baer's Law, finds exceptions in caenogenesis (especially modifications for telolecithy and cleidoic eggs), but is still generally true as regards gross morphology *after* the phyletic stage has been achieved. Eggs within one group of animals, for example the vertebrates (p. 9), may differ very greatly, but they **converge** during early, maternally-programmed development. Later they **diverge** toward their various adults; the females again produce a variety of eggs with a common phyletic programme.

This conservatism has also led to the view that "ontogeny repeats phylogeny". This would only be universally true if all evolution proceeded by **gerontomorphosis,** the adding of new stages *after* the ancestral adult stage. Both heterochrony and caenogenetic characters modify the phylogenetic sequence. If ontogeny repeated phylogeny exactly, then an ancestor of man would have lived on milk all his life and a more remote ancestor would have spent his days attached to his mother by the umbilical cord! Despite this, the statement does have an element of truth and much may be deduced about the ancestry and relationships of an animal species from a careful consideration of its embryology.

APPENDIX 1—SOURCES OF MATERIAL

Introduction

All of the organisms mentioned here are sexual and so can, in theory, be expected to produce observable sperms, eggs and development. However, this is often not so. Only *Pomatoceros* can easily be persuaded to give all these; the other organisms may show development but not gametes, sperms but not eggs, blood systems but not gut, or whatever. Of the vertebrates, the zebra fish or Siamese fighting fish are the most useful (or the mouse, if the course is medically oriented). We have been rude about *Xenopus*, because we find it less satisfactory than the above; in some peoples' hands it has proved adequate (and its tadpoles are beautiful!). For more detailed descriptions of the methods to use for these and many other forms the reader is recommended to Billet and Wild (1975).

(1) Rhabditis

This is a small nematode worm (2 mm male, 5 mm female) whose eggs are found in the soil and in live earthworms. Kill an earthworm by *very brief* immersion in boiling water and drape the body over some soil in a Petri dish or other closeable container with *wet* filter paper in the bottom; it is important to keep the system damp-to-wet; if you are using vented Petri dishes (and most modern disposable ones are), the preparation should be checked and rewetted, if necessary, every day. Nematode worms will be found wriggling over the surface of the decaying earthworm or in the liquid below it in about 3–7 days. If you can bear the smell, a much better yield is produced by the second generation of nematodes about 4–7 days after the first one. We find 18–21 °C a suitable temperature, but experiment with this in case your times are different from ours. To examine female Rhabditis, pick up the largest ones you can see with a lens or stereo microscope (at least 3 mm) with a needle or watchmaker's forceps: transfer them to a tiny drop of water on a microscope slide so that the cover slip compresses them. As the worm is compressed eggs will pop out of the reproductive opening (in the middle of the worm), the first ones hatching as they emerge (avoid particles of sand under the cover slip—they will stop you compressing the nematode). Pressure on the cover slip will squeeze out all the eggs and in every female a selection of cleavage stages can be beautifully displayed. Amoeboid sperms can usually be seen in the area of the unfertilized eggs.

Male *Rhabditis* have two spicules at the rear end for mating.

(2) Pomatoceros

These worms can be obtained from Plymouth Marine Laboratory or other marine stations which supply organisms to universities and schools, at nominal prices, throughout the year. Their great advantage is that they will produce eggs and sperms at any season and fertilization is usually not difficult, even in midsummer or midwinter. The worms should be removed from their tubes *with the minimum of damage to the worms*, as follows. Break off the *thin* (posterior) end of the tube with a metal point and push the worms *down* the tube (out of the *thin* end) with a probe. Males may be recognized by their yellow abdomens and females by their red or violet abdomens. The naked worms should be put in dishes containing a small amount of sea-water, and should be kept cool; egg and sperm release usually occurs within 15 minutes.

The eggs on laying are uniformly pink and the nucleus is clearly visible. Many are normally not round but angular, from compression. Pick them up very carefully with a pipette; *do not use a pipette which has been used for sperms*: Concentrate the egg suspension by holding the pipette vertically in order that the eggs may fall into the narrow tip. The eggs are best observed in just enough sea water to prevent the coverslip from squashing them; close the iris of the microscope or use phase contrast if available. The sperms are concentrated in the white cloud which the males emit; a *tiny* drop of this will spread under a coverslip and keep the sperms in one plane—too much of the suspension makes them difficult to observe. Use phase contrast 400, or very low apertures (condenser iris closed).

To observe fertilization a concentrated drop of eggs and a tiny drop of sperms are placed 5 mm apart on a clean microscope slide. The coverslip is dropped on and they are observed immediately on a microscope *already set up* for this purpose (use some eggs). Where the egg and sperm suspensions overlap fertilizations will be seen in the first 30–60 seconds. You may be lucky enough to see other fertilizations, further into the egg suspension, occurring later as the sperms swim in. At fertilization, the pink pigment moves round to one side of the egg, so that most of it will later be incorporated in cell 4d, which will produce the mesoderm of the adult worm. At the same time the fertilization membrane elevates (often only slightly) and the *nucleus becomes indistinct* as syngamy occurs; the first cleavage then begins. It is usually better to take successive samples from a large dish to see the cleavages rather than to watch one egg on your slide; it is important to keep the container cool.

If you want to make permanent stained preparations of the eggs or embryos you will find it helpful to use gentle centrifugation to collect the eggs after fixation, de- and rehydration and staining.

(3) Helix

In the spring, when the snails have recently mated, the spermatheca contains sperm which are very long, thin and vibratile. A spermatophore can usually be found in the flagellum or sperms can be seen in smears of the ovo-testis. Do not confuse the sperms with the Protozoa (*Crytobia* sp, which resemble trypanosomes) which are usually found in all these sites.

(4) Pond snails (*Limnaea* and *Planorbis*, not *Palludina*)

These snails will usually lay egg masses if first starved for a week and then provided with food (commercial fish food, or lettuce, or *Elodea*). Within the capsules inside each egg mass cleavage stages, trochophores and veligers can be seen; unfortunately all will be at the same stage in one egg mass. Late veligers (with a spiral shell) can be made to undergo premature torsion by illuminating them strongly on a microscope slide.

(5) Echinoderms

The difficulty with *Echinus* and *Asterias* is that ripe eggs and sperm are only present in the animals up to the time of their normal spawning (around or just after Easter in these latitudes). Assuming that the animals have not previously spawned, they can be persuaded to release eggs or sperm by injecting up to 5 ml of 5 per cent potassium chloride (made up in sea water) into one of the arms or the aboral disc. Keep the animal covered in seawater whilst waiting for the gametes to be shed—this may take from 5 to 30 minutes. Alternatively, the gonads themselves may be dissected from the animals and ruptured in sea water. Examine eggs and sperms as described for *Pomatoceros*. Fertilization can also be observed fairly easily, but keep the sea water container cool if you wish to examine cleavage and subsequent development (avoid having too thick a suspension of eggs and sperm in the container; keep it cool in a refrigerator).

(6) Dogfish

Dogfish egg cases ("Mermaids purses") may sometimes be found on the shore or in tanks in which dogfish have been kept. It is worth bringing them into the laboratory in aerated sea water, keeping them cool and following the development of the young fish.

(7) Trout

The eggs are not very suitable for the study of development, even though they are easy to obtain. They are too large to be looked at easily under a microscope and too small to be examined without one. The ease of keeping them alive (under a running tap) is countered by their taking 6 weeks to hatch. The tropical fishes described below are in all respects more useful.

(8) Tropical aquarium fishes

(a) Live-bearing tropical fishes of the family *Cyprinodontidae:* the best and most common fishes of this family are the guppy, the platy and the swordtail (nominal prices at most good pet shops). The females are in a perpetual state of pregnancy and in the latter two species **superfetation** often occurs, that is to say groups of embryos of various stages are to be found in one female. With a little experience the state of development of the embryos can be assessed by the bulk of the female; her head should be cut off neatly, and her abdomen opened. The ovary is always clearly recognizable by the clear yellow of the yolk, or the eyes of the babies. The eggs should be separated with forceps and those with eyes may be artifically "hatched". Fertilization is impossible to observe in these forms.

(b) Fishes of the family *Anabantidae* (*Osphronemidae*): the best and most common examples, found in most pet shops which keep fish, are the Siamese fighting fishes, the paradise fishes *Macropodus opercularis* and the gouramis, especially *Trichogaster trichopterus*, the blue or three-spotted gouramis and the dwarf gourami, *Colisa lalia*. Pairs should be kept separated at about 26 °C until required for breeding, then placed together at 28–30 °C without other fish in the tank, but with plant cover for the female, who is often attacked viciously. Courtship should commence immediately and the eggs are normally laid about 36–48 hours later (each pair of fish will have a constant interval at a given temperature, and egg-laying can be predicted to the hour in occasional cases). The males of these species blow a bubble nest under which the female is enticed (dwarf gouramies need floating plants, or cottonwool, to stabilize the nest). The male wraps his body round hers and squeezes out the eggs while releasing sperm. The eggs are then put into the nest by the male, whence they can be removed by the callous human with a beaker. They develop quite well in very shallow water (<0.5 cm) at 25–30 °C, but are very sensitive to temperature change. Early cleavages occur every 15 minutes, and they hatch in about 2 days.

The sperms, which are very small, may be collected by squeezing the male *gently*—or you may find active sperm in the nest during mating.

(c) The zebra fish, *Brachydanio rerio*, is probably the most reliable and convenient source of developing vertebrate embryos. The fish are small, reaching a maximum of under 5 cm, so quite a large shoal may be kept by even a modest establishment. A tank (with a few floating plants that can be removed while catching the fish and then replaced or with plants weighted by lead strip) whose temperature can be maintained between 24–28 °C, will do admirably. Tap water may be used for setting up, if it is not too hard or soft, otherwise lake or stream water (*without* dirt, which may also carry parasites) may be used. A 40 l (30×30×60 cm) tank will comfortably accommodate fifty or more adult fish. Stock may be purchased at any pet shop, usually as immature fish, at nominal prices; they should be fed some live food or, failing this, scraps of proprietary cat or dog foods, as well as dried fish foods. Feed sparingly, as with all fish, and siphon off sediment from the bottom of the tank periodically (do *not* use gravel but a very thin layer of fine sand or leave the bottom bare). To obtain the eggs, first catch two plump females (silvery with even stripes) and three or four males (yellowish tinge and broken stripes above the anal fin) and place them in the breeding container with as little handling as possible. The breeding container may be another tank with a metal or plastic mesh basket hanging in it, as wide and long as possible (more than 15×30 cm) but *not* more than 5–8 cm deep, hanging in the water. Or the breeding container may be a large shallow polythene tray, with a layer of marbles covering the bottom. The container should be filled with water from the stock tank and its temperature must be maintained at about 28 °C; it may be stood in a large waterbath (avoid vibration), or even over a radiator, but it is not practicable to heat *shallow* water with a standard aquarium heater and thermostat. Best to float it in a larger aquarium, or to put it in a warm *humid* greenhouse.

During the night the fish will spawn, scattering non-adhesive eggs about 1 mm diameter, which they will eat if they can! In the breeding tank, the eggs will fall through the mesh or between the marbles, from which they may be siphoned. Two females should produce about 80 eggs the first night, and may spawn on the following night, but it is better to leave them for about 2 weeks before attempting to breed from them again. In some circumstances, the fish spawn at first light, and if their tank is darkened will often spawn when the cover is removed. Sometimes the fish will even spawn in a bowl with a layer of marbles lowered into the stock tank. The eggs are beautifully transparent (Plate IX) and the time sequence of development is approximately:

Fertilization membrane	2–3 min
Aggregation of blastoderm	10 min
First cleavage	20–30 min
Second cleavage	40–60 min
Subsequent cleavages—every	15--20 min
Subgerminal cavity	4–5 h
Gastrulation commences	5–6 h
Gastrulation completed	8–14 h
Eyes	12–16 h
Heart beating	20–25 h
Pigment in eyes, and melanocytes	30–36 h
Hatching	70–85 h or less

The young fish should be fed sparingly on boiled egg yolk, then newly hatched rinsed brine shrimp, then standard fine fish foods. They are mature in 4–6 months. The older fish will eat them if they can do so; therefore, delay introducing young fish into the stock tank.

(9) Rana

Frogs will often go into amplexus, etc., from January through to June, if they are kept in the refrigerator in a small amount of water and then brought into the lab., which is usually warmer, with rather more water. The eggs are usually laid 4 or 5 days later. Frog spawn is simple to keep in the laboratory and the ease with which later stages of development can be observed, including the classic vertebrate metamorphosis, make it a star performer in a study of living embryos. Early stages of development, immediately after fertilization, are not easy to see, however, because of the mass of dark pigment on the cells.

Developing tadpoles may be fed with powdered nettle or even tiny bits of beef heart hung on bits of string; during metamorphosis, they will readily take *Tubifex* worms and other small invertebrates. Do remember that baby frogs must be allowed access to air (put in a stone or something similar that they can climb on to).

(10) Newts

Newts, too, will breed if spring is imitated in the same way as for frogs.

(11) Axolotls

Axolotls, conversely, will often breed after acute cooling—drop ice-cubes into the tank to take the temperature down towards 0 °C! Their neurulation occurs on the third day after laying and is beautiful, probably easier to see and understand than that of any other animal.

(12) Xenopus

This has now become the standard reproductive organism for British schools, which seems a pity: it leads many children to believe that sex depends upon being injected, that a grotesque amplexus is on the route towards an elegant copulation and that the study of embryology is exemplified by opaque eggs, many of which develop abnormally. Despite these reservations, *Xenopus* are easy to keep and can be made to spawn at any time of the year.

Mature pairs should be kept in tanks about 20×20×45 cm, about one-third full of water, *covered securely*. Pairs which have not bred in the past month and in which the female looks fat (feed beef heart and liver) should be injected with hormones as follows: hold the animals with a wet tea-towel (*not* starched); injection should be into the dorsal lymph space above the pelvic girdle; the hypodermic needle (21G 3 cm) is inserted under the loose dorsal leg skin and gently pushed up between the muscles into the loose dorsal skin above the "hip"; injection should take about 30 seconds; a priming dose of 50 i.u. of human chorionic gonadotropin (HCG) for male, 100 i.u. for the female and then an ovulating dose of twice this in each case, 24 hours later; egg laying should occur on the morning of the third day.

(13) Fowl

Eggs must be certified *fertile* when bought—those from the grocery store are useless for embryology! They should be incubated in a humid atmosphere at 38–39 °C and turned each day for the first week of incubation. They may be stored in a cool place (6–10 °C) for about a week without development or deterioration, but only *before* they are incubated. Take from the incubator only those eggs needed for a particular class at one time and do not chill the others for long periods (although moderate chill for short periods does little harm).

(14) Mammals

For the most part this section will deal with mice. However, mammalian sperms are easy to obtain, especially from the human. They come in a variety of shapes and those of rodents are especially dramatic with their hooked heads. Unfortunately, rodents must be killed and sperms squeezed from the *cauda epididymis* or *vas deferens*, *not* from the testis. Use warm mammalian saline in which to collect and suspend them and warm the slide if normal motion is to be observed.

The mouse is the most convenient mammal from which to produce early eggs and embryos. Although it is relatively easy to induce female mice to ovulate by injecting them intraperitoneally with 2 i.u. of follicle-stimulating hormone (FSH, usually as pregnant mare serum), followed 45 hours later by 2 i.u. of HCG, causing ovulation 12 hours later, it is probably easier to use natural mating. Keep about four females with a male and examine them each morning for a vaginal plug. First cleavage occurs late on the day of the plug (day 1), second cleavage on day 2, third on day 3, blastocyst on day 5–6 and implantation occurs around day 6. Cleaving eggs can best be viewed while in the fallopian tube by removing the tube (tiny tube forming a knot right next to the ovary) to a slide in mammal saline, unravelling it with a couple of pins and then pressing a cover slip on moderately hard. The ciliary beat in the tube is very impressive too. Later embryos may be used for classes to show membranes and are useful to see the developing organ systems "in the round" (you need a good stereo microscope).

APPENDIX 2—METHODS

(1) The Vital Staining of Embryos

This is intended to demonstrate the flow of tissues during the very early stages of development; the technique can be acquired very quickly and is really very successful.

(a) *The preparation of staining rods*

(i) Small glass rods should be drawn out to a point rather finer than a hair and a small (*just visible*) blob should be run on the end of them.

(ii) These blobs should now be dipped into warm agar (or gelatin or egg albumen and then fixed in alcohol or formaldehyde).

(iii) Staining of the tips with Nile Blue, Methylene Blue, Crystal Violet or Neutral Red is performed either by dipping the rod into a strong solution of the stain overnight (preferably in alcohol) or by sticking a tiny crystal of the stain to the tip and then covering this with agar (or albumen or gelatin, which is then fixed in alcohol or formalin).

(iv) These staining rods may best be preserved in dry tubes. Wash in appropriate saline before use or they stick to tissues.

(b) *The preparation of the embryo*

(i) Either very young (blastoderm) states of the live-bearing fishes or 12–24-hour chick embryos should be used.

(ii) The embryo is removed either from mother or egg-shell and placed under a good stereo microscope in appropriate saline.

(iii) If it is desired that the fertilization membrane should be removed, saline should come about two-thirds up the egg, the upper part of the egg thus being left dry. The membrane over this will crinkle and can easily be lifted off with forceps. The embryo must, of course, then be covered with saline.

(iv) If the fertilization membrane has not been removed quite a dark stain mark may be made over the edge of the blastoderm of fishes, or over the primitive streak of the 6–12-hour incubated chick. Stain the vitelline membrane, and the tissue (it is lightly stained, often in quite a different colour) may be seen to move from under the mark on the membrane. If the membrane has been removed then a much paler mark should be made, and it is often possible to make several small marks and thus to work out the pattern of movement in its entirety. Usually the heat from the lamp illuminating the embryos from above is sufficient to keep the tissues alive. If this experiment is attempted with trout eggs, the heat will kill them.

(v) In our Zoology Department we have successfully hatched fish eggs which have been maltreated in the way described and have kept chick eggs 48 hours with apparently normal development. It is important, of course, to top up the solution with *distilled water* to make up losses due to evaporation.

(2) The Systems of Older Embryos

Apart from the time-honoured method of displaying selected sections of the particular "stages" (which imposes upon both teacher and pupil a somewhat unnecessary burden of three-dimensional imagination) there are several other techniques which may be usefully employed either to supplement or to replace this.

(a) The Blood Vascular System

This may be seen most easily in living embryos or very early stages (post-hatching) of fishes. These have the additional advantage of showing a typical simple vertebrate blood vascular system in diagrammatic form. They also give the student a sense of the dynamics of blood flow which he does not get from the frog's foot.

(b) The Urinogenital System

Dissection of chick embryos at 72–96 hours of incubation is very simple under a stereo-microscope and, if the gut can be removed intact, shows pronephros, mesonephros and the various ducts lying in the dorsal peritoneal wall very beautifully, especially if lightly stained with methylene blue or crystal violet.

(c) Tissue Culture Methods

These may usefully be employed by those with some experience of the techniques, and frequently give considerable help in the teaching of organogeny.

APPENDIX 3—FURTHER READING

There are few, if any, textbooks of Embryology which treat in more detailed manner all the topics touched on in this book. Perhaps the best general textbook is:

BALINSKY, B, I., *An Introduction to Embryology*, 4th ed., 1975.

More encyclopaedic, and with most emphasis on vertebrate work:

NELSON, I. E., *Comparative Embryology of the Vertebrates*, Blakeston, 1953.

For general reproductive topics and a rather fuller treatment of the theoretical background to development:

COHEN, J., *Reproduction*, Butterworth, 1977.

With a much more functional approach and less basic embryology:

EDE, D. A., *An Introduction to Developmental Biology*, Blackie, 1978.

A very good laboratory manual for experimental embryology of the vertebrates is:

BILLET, F. S. and WILD, A. E., *Practical Studies of Animal Development*, Chapman & Hall, 1975.

A very comprehensive guide to human embryology is:

HAMILTON, W. J., BOYD, J. D., and MOSSMAN, H. W., *Human Embryology*, Heffer, Cambridge, 3rd ed., 1962.

Two slightly less comprehensive, but very easy to follow, human embryology books (they are widely used by Birmingham Medical Students) are:

LANGMAN, J., *Medical Embryology*, Williams & Wilkins Co., 1975.

MOORE, K. L., *The Developing Human*, Saunders, 1977.

For straight descriptive organogenesis, there are few modern books as good as:

GOODRICH, E. S., *Studies on the Structure and Development of Vertebrates*, New ed., Dover, 1958. (Original publication 1930.)

For a review of genetic control of early development: DAVIDSON, E. H., *Gene Activity in Early Development*, Academic Press, 1976.

For a very elegant, readable, yet scholarly discussion of the relationship of embryology and evolution:

GOULD, S. J., *Ontogeny and Phylogeny*, Harvard University Press, 1977.

GLOSSARY

Animalculists: philosophers who believed that the sperm contained the homunculus, or at least all the necessary information to make the embryo.

Ankylosing spondylitis: a human pathological condition in which the vertebral discs fuse together; much commoner in people with HLA B—27.

Antibodies: proteins (globulins), produced by special cells of an animal, which can form a specific, strong association with another molecule, usually of foreign origin, called the antigen.

Antigen: see **Antibodies.**

Blastoderm: the layer of cells in a blastula which surround the blastocoel; the layer of comparable cells in a telolecithal embryo.

Blueprint: a precise set of instructions as to the way something has to be made, by *describing* the finished product, as in engineering drawings (which used to be on blue paper). Contrast *prescription* which gives only general directions for making something.

Chloroplast: cellular organelles found in plants which contain chlorophyll as well as the other chemicals associated with photosynthesis. Probably originally a symbiont, as they contain their own DNA rings.

Chordal plane: a plane parallel to a surface (or nearly so). In a sphere, chordal planes are at 90° to radii.

Clone: the progeny of but one cell or organism, derived by mitosis and therefore presumably genetically identical.

Crossovers: the process whereby genetic material is exchanged between two chromatids when they come together during meiotic prophase. It is possible that the exchange of material between two chromatids is often not equal, one donating a larger segment to the other than it receives; it is assumed that this *may* lead to a "nonsense" message on both resulting chromatids.

Delamination: the separation of successive layers (laminae).

Diploid: see **Polyploid.**

Ecto-mesenchyme: see **Mesenchyme.**

Embryonic axis: the important axis of an embryo, e.g. the head-tail axis of vertebrates; *axial organs* are nerve cord, notochord and usually gut.

Embryonic disc: the disc-shaped mass of cytoplasm, or later on a mass of cells, which is the early embryo of some telolecithal eggs, e.g. chick.

Epithelium: a complete layer of cells covering (or forming) a surface in an animal or plant.

Extirpation: destruction and/or removal of an embryonic structure, usually to investigate its effects on other structures or tissues.

Genotype: the array of genes possessed by an organism; also called the **genome.**

Germ cells: those cells whose progeny will give rise to the gametes of an organism. In most developing metazoan embryos the germ cells are determined earlier than any other cells.

Homeotherm: an animal which maintains a constant body temperature; contrast poikilotherm, which does not.

Homograft: a tissue graft between animals of the same species; contrast allograft, from another species; autograft, from the same animal.

Haploid: see **Polyploid.**

Histones: proteins, rich in the amino acid lysine, found in chromosomes. These proteins are assiocated with DNA and are probably involved in switching off gene function.

Homologous: used to describe similar organs in two organisms if they are supposed to derive from one organ in the common ancestor; e.g. arm of man, wing of bat and bird, but not of insect.

Homunculus: the "little man" believed by some philosophers to be contained in gametes, requiring only to grow to produce the adult philosopher; why it turned into a baby on the way is unclear.

Lappets: folds of material or tissue, resembling earlobes or pieces of plasticine pulled out by thumb and finger from a lump.

Lecithin: the major fatty component of yolk.

Ligament: strictly speaking a collaginous binding tissue between skeletal elements (contrast tendon between muscle and skeleton). The term is misused in embryology to refer to the reduced mesenteries supporting peritoneal organs, like the falciform ligament of liver or the broad ligament of uterus.

Lipases: enzymes which break down fats.

Luminal: referring to the cavity (lumen) of organs.

Meiosis: the process of cell division which results in halving the chromosome complement in the cells produced by the division. The commonest process in which meiosis is involved is in the production of gametes, during which the diploid set of chromosomes found in most somatic cells is reduced to a haploid set in the functional gametes.

Mendelian ratios: the ratios of different parental characteristics observed in a collection of offspring. Mendel showed that in many circumstances the ratios suggested that the genes controlling the parental characteristics were randomly assorted into gametes which then fertilized without bias; i.e. no "mixing" or "dilution" seemed to occur.

Mesenchyme: cells of pre-connective tissue, still in a loose association before the formation of ground substance (e.g. in cartilage) or fibres (e.g. collagen).

Mitochondrion: the organelle found in the cytoplasm of eukaryotes which is responsible for the great majority of oxidation reactions associated with the production of energy for the cell. The mitochondrion is probably a symbiotic organism in origin, as it has its own DNA ring.

Mitosis: the normal process of cell division in eukaryotes, as a result of which the number of chromosomes in the nuclei of the daughter cells is the same as in the parent cell; contrast meiosis in which the chromosome number is halved.

m-RNA: messenger RNA, used by ribosomes to make proteins via the genetic code; its sequence of nucleotides is the "negative" of a longer part of the DNA sequence of the organism.

Nucleic acids: the DNA and RNA of the nucleus and cytoplasm. Nucleic acids consist of chains of nucleosides (which contain ribose or deoxyribose sugars as well as an organic base) bound into chains by means of phosphate links between the sugars.

Ovulation: the process of egg release from the ovary.

Patent: having a continuous hole, usually through the middle of a tube; contrast blocked.

Phagocytosis, phagocytose, phagocytize: engulfment of particles by cells.

Polyploid: the condition in which the chromosome complement of the nucleus is found in multiple complete (euploid) sets; compare haploid (one set per nucleus), diploid (two sets), tetraploid, etc.; contrast aneuploid.

Primitive: an animal supposedly resembling the early "simple" ancestor of any particular animal group.

Primordium: the earliest identifiable cell aggregation, whose product is a particular organ, is the primordium of that organ.

Pro-nucleus: the name given to the nuclear components in the zygote which came from the gametes. After fertilization the pronuclei of the two gametes fuse to form the nucleus of the zygote.

Prospective: referring to the supposed future; e.g. the boy is a prospective man.

Protamine: proteins resembling histones but with much arginine instead of much lysine; characteristic of condensed sperm chromosomes.

Protease: an enzyme which functions to break down proteins.

Rickettsia: symbiotic or parasitic organisms larger than viruses and smaller than most bacteria.

Senility: dramatic increase in the likelihood of death associated with breakdowns in physiological and even anatomical processes.

Specialized: an organism or tissue which has special properties, apparently for one particular task, as a result of which it has lost general capabilities.

Supernumerary: extra.

Symbionts: organisms which live together to their mutual advantage.

Syncytium: an organism or area within an organism in which cell membranes are absent between adjacent nuclei.

Uterine epithelium: the internal lining of the uterus.

Tubulin: the contractile protein associated with mitotic spindles and flagella, equivalent to myosin in muscle.

Vesicle: a small bladder.

Zygote: the product of the fusion of the gametes (usually two). Thus sperm and egg fuse to produce the fertilized egg, the zygote, in metazoans.

SUBJECT INDEX